CASE STUDIES IN INFRASTRUCTURE DELIVERY

THE KLUWER INTERNATIONAL SERIES IN INFRASTRUCTURE SYSTEMS: DELIVERY AND FINANCE

Consulting Editor

John B. Miller, Ph.D.
Massachusetts Institute of Technology
Cambridge, USA

Case Studies in Infrastructure Delivery

By

John B. Miller

Massachusetts Institute of Technology

KLUWER ACADEMIC PUBLISHERS
Boston/Dordrecht/London

Distributors for North, Central and South America:
Kluwer Academic Publishers
101 Philip Drive
Assinippi Park
Norwell, Massachusetts 02061 USA
Telephone (781) 871-6600   Fax (781) 681-9045
E-Mail < kluwer@wkap.com>

Distributors for all other countries:
Kluwer Academic Publishers Group
Distribution Centre
Post Office Box 322
3300 AH Dordrecht, THE NETHERLANDS
Telephone 31 78 6392 392   Fax 31 78 6546 474
E-Mail < services@wkap.nl>

 Electronic Services < http://www.wkap.nl>

Library of Congress Cataloging-in-Publication Data

Miller, John B.
    Case studies in infrastructure delivery / by John B. Miller.
    p.cm.—(The Kluwer international series in infrastructure systems. Delivery and Finance:
    ISDF 102)
    Companion text to the author's Principles of public and private infrastructure delivery.
    Includes bibliographical references.
    ISBN 0-7923-7652-8 (alk. paper)
    1. Infrastructure (Economics)—United States—Finance—Case studies. I. Title. II.
    Series.

HC110.C3 M527 2001
363'.0068'1--dc 21
                                                                                          2001058518

Excel® is a registered trademark of Microsoft Corporation. Adobe®
and Acrobat® are registered trademarks of Adobe Systems
Incorporated.

Printed on acid-free paper.

Printed in the United States of America.

TABLE OF CONTENTS

LIST OF TABLES

LIST OF FIGURES

LIST OF EXHIBITS (STORED ON COMPANION CD)

FOREWORD

This book of case studies is the companion text to *Principles of Public and Private Infrastructure Delivery*, published by Kluwer Academic Publishers in October, 2000. Together, these are intended to provide an in-depth look at how project evaluation, procurement, and packaging are revolutionizing the role of civil engineers in managing collections of infrastructure assets. Although these texts were developed for one of my courses at MIT, the set offers wide ranging opportunities to introduce students, at both the undergraduate and graduate levels, to very broad, yet specific definitions of project evaluation. Each text introduces students to the astounding role engineers, constructors, financiers, and inventors have always played in American public infrastructure development.

These texts provide a new look at civil engineering project evaluation – revitalized for the 21st Century. A new approach to civil engineering project evaluation recognizes that in the broadest sense, civil engineering encompasses the broad range of financing alternatives, project delivery alternatives, technology assessment, and environmental analysis that now comprise good civil engineering practice. The cases studies acquaint civil engineers, their Clients, and society generally, to the extraordinary power of a simple analytical framework, reasonably accurate project information, and good problem solving skills in the development of workable alternative solutions to complex public and private capital programming problems.

The Quadrant Framework for Analysis of Project Delivery and Finance Alternatives

A key analytical tool from the *Principles* text is a Quadrant Framework that is used throughout the book as a means of comparing and contrasting different delivery and finance strategies for infrastructure projects and services. To facilitate ready use of this text by students who may not have the Principles text handy, the Quadrant Framework is also described here. (See, Chapter 2 in the *Principles* text for the complete presentation).

The purpose of the framework shown in Figure 1 is to compare and contrast infrastructure funding and delivery patterns. The framework is not a mathematical model. Rather, it is a practical tool for configuring stable, sustainable collections of infrastructure systems and facilities. The framework is also explored in depth in Chapter 6 of the *Principles* text, where it is used to help infrastructure managers build alternative scenarios for local, regional, and national infrastructure portfolios.

Based upon the data collected, trial metrics were applied to see if a framework could be created in which project delivery and project finance methods were distinguished and compared as interdependent variables. The metric selected for project delivery methods was the degree to which typical project elements are separated from each other. This metric was applied from the viewpoint of the CLIENT, that is, at the interface between the CLIENT and third parties. For example, in a typical Design-Bid-Build project, key elements of the project -- design, construction, and operations/maintenance -- are provided by separate, independent participants. On the other hand, in a typical Design-Build-Operate franchise for wastewater treatment, all these project elements are combined.

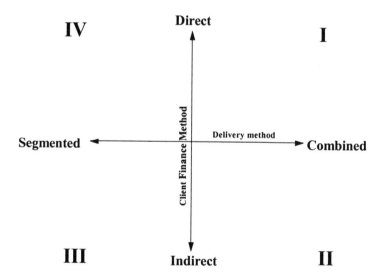

Figure 1 The Quadrant Framework

The metric selected for project finance methods was the degree to which government assumes the direct financial risk for producing the project. For example, government assumes all the financial obligations, including progress payments during construction, in the typical Design-Bid-Build format. On the other hand, government assumes no direct obligation for capital costs, operations costs, or financing costs in a typical Design-Build-Finance-Operate franchise in Hong Kong. The DBFO franchise award, through which the government arranges for the collection of a revenue stream by the franchisee, is an indirect means of encouraging the franchisee to assume these risks. Variations in financing between these two extremes are a continuum in which financial responsibility is shifted between government and private sector sponsors.

Comprised of two perpendicular axes, the horizontal axis represents the continuum of delivery methods measured by the degree to which typical elements are segmented or combined with one another, while the vertical axis represents the continuum of financing methods measured by the degree to which government assumes the financial risk for producing, operating, and maintaining the project throughout its life cycle.

The framework uses these two axes to describe government's two fundamental strategies for promoting infrastructure development. Each axis represents a continuum of choices for government, one in the area of funding and the second in the area of project delivery.

The first strategy -- the vertical axis -- describes the range of potential choices to be made by government to fund (or finance) infrastructure. The vertical axis depicts a strategic election by government to arrange project funding somewhere between two fundamentally different extremes:

- (a) to "push" specific projects "directly" through current cash appropriations, or

- (b) to "pull" specific projects "indirectly" through incentives, mandates, dedicated income streams, or other measures, which encourage the private sector to finance Government goals.

Current appropriations are typically used for specific contracts or grants funded directly by the government. For purposes of the framework, funding is considered to be "direct" if government is the source of cash to finance the design and construction of a particular infrastructure project. The significant administrative requirements of federal grant programs, which tie state and local government to federal acquisitions standards, are considered to be "direct" federal funding.

The second strategy -- the horizontal axis in Figure 1 -- describes the range of potential choices to be made by government for project delivery. The horizontal axis represents a strategic election by Government to approach planning, design, construction, operations, and maintenance, in one of two, fundamentally different ways:

- (a) by clearly separating each of these different steps in the procurement process from one another (a "segmented" process) or

- (b) by combining all these aspects of an infrastructure project into a single procurement of the completed facility (a "combined" process).

With these two fundamental strategies arrayed on the axes in Figure 1, infrastructure development strategies can be graphically described and compared both on program-wide and project specific bases. The quadrants defined by these two axes are numbered for convenience.

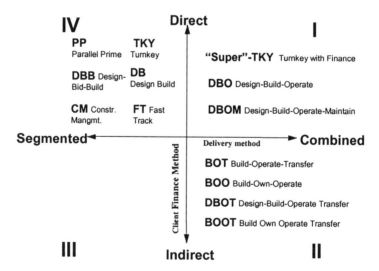

Figure 2 Fitting the Project Delivery Methods into the Four Quadrants

Figure 2 summarizes how the common project delivery methods fit into the Operational Framework. Several variations of Design-Build-Operate and Design-Build-Finance-Operate are added. The assignment of delivery method to quadrants is described in more detail above.

These case studies drive home the view that infrastructure development and renewal is never "over," never finished. The American experience confirms that infrastructure development and economic development are intimately related, and that with a bit of planning, sound financial analysis, and tough competition, governments and private sector firms can restore stable procurement systems to make wise infrastructure renewal and improvement   an   integral   part   of   economic   development.

PREFACE

The purpose of these case studies is to provide students of all types – undergraduate, graduate, executive, and citizen – appropriate knowledge of how infrastructure services and projects can be effectively delivered and financed in the modern world.

The term "delivery" is used to describe the continuum of project delivery methods regularly applied by the construction industry and taught by academic programs in project management and construction engineering throughout the world. These delivery methods include sequential Design-then-Build (Design-Bid-Build, Design-Build, Parallel Prime, Fast Track, Construction Management (at risk), Construction Management (not at risk), Turnkey, Design-Build-Operate, Build-Operate-Transfer, Design-Build-Finance-Operate, and Build-Own-Operate.

The term "finance" is used here to describe the continuum of approaches applied to infrastructure project finance by public and private owners, quasi-public agencies, developers, constructors, financiers, bankers, investment bankers, and fund managers. These approaches include complex combinations of public and private sector debt and equity; sovereign obligations, commitments, statutes, regulations, and other incentives, along with private sector guarantees, commitments, and assurances. These approaches are the subject of academic programs in management and finance throughout the world. While this financial expertise is also of great importance in the implementation of individual infrastructure projects, it is not the focus here.

The companion text, *Principles of Public and Private Infrastructure Delivery*, published by Kluwer Academic Publishers in October, 2000 treats "delivery" and "finance" of infrastructure as two separate, but essential, elements of emerging strategies for competitive infrastructure development.. The optimal combination of these two elements, both at the project and system levels, is a central theme throughout these case studies.

The case studies raise numerous questions – frequently involving complex tradeoffs among competing values and goals. These difficulties are an unfortunate, but unavoidable characteristic of large infrastructure systems. The cases point to a new approach for evaluating public infrastructure spending alternatives at the portfolio level. The objective of this approach is to produce a complete, sustainable set of infrastructure facilities at the portfolio level, based upon a series of optimal matches of delivery and finance methods at the project level.

ACKNOWLEDGEMENTS

Many of my Research Assistants over the past six years have been heavily involved with me in the development of these cases. Their contribution is gratefully acknowledged (and noted at the beginning of each case). Over the past three years, my MIT Teaching Assistant, Michael J. Garvin, has led the effort to improve each case based on the experience of MIT student users. Dr. Garvin has been an important contributor to the cases presented here. We have been very lucky to have extraordinary assistance from a number of the principals involved in these projects, many of whom were kind enough to present and discuss the case studies in person at MIT. These special contributions are very much appreciated.

The very special support of the National Science Foundation, through a CAREERS grant to the author, provided important support for students and faculty to keep the Infrastructure Case Study Development effort moving over the past three years. The support of the Ferry Foundation was also instrumental in bringing many of these case studies to life. The National Science Foundation's CAREER grant program provided key support to me, my students, and my staff, for which I am very grateful.

Thanks to MIT's Provost Robert B. Brown for a teaching leave in the Fall of 2000, which provided precious time to complete this second book in Kluwer's *International Series in Infrastructure Systems: Delivery and Finance*. Thanks also to MIT Professors Fred Moavenzadeh, Dan Roos, David Marks, Joe Sussman, Sarah Slaughter, and Rafael Bras for their support. Many thanks also to Jacquelynn E. Henke for effectively managing the conversion of the individual case studies into this manuscript.

Joan, John, Doug, and Mary continue to make very special contributions.

This book is dedicated to Henry L. Michel, P.E., NAE. His many contributions to the world's construction industry and to the education of MIT students live on.

GLOSSARY OF TERMS

AEC Sector ("Architect/Engineer/Constructor"). The sector of the economy which comprises the activities of architectural, engineering, and construction firms engaged in the design, construction, and rehabilitation of constructed facilities, typically buildings, and including, at any tier, specialty engineering firms (electrical, mechanical, civil, geo-technical, environmental, etc.), general contractors, and subcontractors.

Build-Operate-Transfer (BOT). See Design-Build-Finance-Operate. BUILD-OPERATE-TRANSFER (BOT) is defined as a delivery method in which the CLIENT procures design, construction, financing, maintenance, and operation of the facility as an integrated whole from a single PRODUCER. The Client provides initial planning and functional design. As defined here, the BOT method puts the risk that project receipts will be insufficient to cover <u>all</u> project costs and debt service squarely on the PRODUCER.

Client/Owner. The terms OWNER and CLIENT refer to the public or private client procuring facilities or services.

Contractor/Producer. The terms CONTRACTOR and PRODUCER refer to the successful bidder or proposer that emerges as the winner of the procurement process.

Design-Bid-Build (DBB). DESIGN-BID-BUILD (DBB) is defined to mean a segmented delivery strategy in which design is fully separated from construction, both of which are, in turn, separated from maintenance and operation of the facility. In the DBB model, the Client also separately provides planning and financing of the project.

Design-Build (DB). DESIGN-BUILD (DB) is defined as a delivery strategy in which the CLIENT procures both design and construction from a single PRODUCER. Initial planning, functional design, financing, maintenance, and operation of the facility remain as separate, segmented elements of the project, each of which is provided by the CLIENT.

Design-Build-Finance-Operate (DBFO). See, Build-Operate-Transfer (BOT), a commonly used delivery method synonymous with DBFO. DESIGN-BUILD-FINANCE-OPERATE (DBFO) is defined as a delivery method in which the CLIENT procures design, construction, financing, maintenance, and operation of the facility as an integrated whole from a single PRODUCER. The Client provides only initial planning and functional design. As defined here, the DBFO method puts the risk that project receipts will be insufficient to cover <u>all</u> project costs and debt service squarely on the PRODUCER.

Design-Build-Operate (DBO). DESIGN-BUILD-OPERATE (DBO) is defined as a delivery method in which the Client procures design, construction, maintenance, and operation of the project from a single PRODUCER. The Client provides initial planning and functional design. The DBO procurement method is defined to require that the CLIENT directly provide some portion of the cash flows required by the PRODUCER to finance all of the tasks assigned by the CLIENT. This financing is typically provided in one of two ways (and sometimes as a combination of the two): (a) direct cash payments by the CLIENT, or (b) delivery by the CLIENT of the equivalent of direct cash payments to the PRODUCER, such as the right to collect user charges. Even if the successful PRODUCER is required to provide some portion of the funding stream required for project delivery or operation, such projects are classified as DESIGN-BUILD-OPERATE (DBO).

Discounted Cash Flow. Discounted life cycle cash flow models provide the basic building blocks from which comparative analyses of alternate configurations of infrastructure portfolios are made. As CLIENT and PRODUCER sign each contract, the parties make a mutual commitment to exchange services, supplies, and equipment for an agreed cash flow stream. The elements required to be delivered by the PRODUCER and the actual cash flows committed by the Client are fully described in the contract. This contract cash flow is used to represent each project and each delivery method in the portfolio. Cash flow over a common planning period is the common denominator for comparing the effect of different project delivery methods on cumulative cash flow of each project, and collections of projects, in the client's portfolio.

EPC Sector ("Engineering Procurement Construction"). The sector of the economy comprising the ("E") engineering, ("P") procurement of materials and equipment, and ("C") construction of constructed facilities, typically industrial plants and large public works facilities. Typical examples of projects in this sector include power, refining, chemical, and manufacturing plants, and water, wastewater, and transportation facilities.

Infrastructure. The term "infrastructure" is used in a broad sense to mean, collectively, (a) capital facilities such as buildings, housing, factories, and other structures which provide shelter; (b) the transportation of people, goods, and information; (c) the provision of public services and utilities such as water; power; waste removal, minimization, and control; and (d) environmental restoration.

Owner/Client. The terms OWNER and CLIENT refer to the public or private client procuring facilities or services.

Portfolio of Projects. The term "portfolio of projects" is used to refer to a collection of infrastructure facilities and services owned, leased, operated, or controlled by a single CLIENT.

Producer/Contractor. The terms CONTRACTOR and PRODUCER refer to the successful bidder or proposer that emerges as the winner of a procurement process.

Project Viability. Project viability is defined as the combination of technical, financial, and environmental feasibility, the key ingredient for effective use of all the project delivery methods. For example, a proposed waste water treatment project that generates sufficient revenues from sewer use fees to pay for (a) maintenance and operations through its useful life, (b) debt service on the cost of initial construction, and (c) a reasonable profit to PRODUCERS along the way can be delivered using any of the delivery methods in common use throughout the world.

Project. The term "project" is used to refer generically to discrete tasks performed in connection with part, or all, of an infrastructure facility, or service. Projects are often arranged to be performed by private or public owners ("CLIENTS") through contracts awarded to suppliers, designers, contractors, design-builders, design-build-operators, operators ("PRODUCERS") for all or part of a capital facility, repair or replacement of components of such facilities, or for services relating to infrastructure facilities.

Quadrant I. The portion of the Quadrant Framework that is defined by combined project delivery methods and direct project finance methods.

Quadrant II. The portion of the Quadrant Framework that is defined by combined project delivery methods and indirect project finance methods.

Quadrant III. The portion of the Quadrant Framework that is defined by segmented project delivery methods and indirect project finance methods (generally unused, but see, Appendix C for a description of Superfund).

Quadrant IV. The portion of the Quadrant Framework that is defined by segmented project delivery methods and direct project finance methods.

CHAPTER 1   BRIDGING THE GOLDEN GATE: OUTSOURCING TO A NEW PUBLIC ENTITY

Infrastructure Development Systems IDS-00-T-016

Research Assistants Bradley Moriarty and Kai Wang prepared this case under the supervision of Professor John B. Miller as the basis for class discussion, and not to illustrate either effective or ineffective handling of infrastructure development related issues. Data presented in the case has been altered to simplify, focus, and to preserve individual confidentiality. The assistance of the Golden Gate Bridge, Highway and Transportation District is gratefully acknowledged. To learn more about the project, the District's publication "The Golden Gate Bridge: Report of the Chief Engineer" is an invaluable resource.

Bridging the Gap

San Francisco thrived throughout the 1800's as a popular and busy port. With the arrival of the trans-continental railroad in Oakland in the 1860's, San Franciscans watched port related commerce spread out of their city toward Oakland. The only rail access to San Francisco was by the Dumbarton crossing far to the south. Although San Francisco's importance as a financial center continued, talk about establishing a fixed crossing of the bay to San Francisco led nowhere.

In the 1900's, the automobile created the rise of fixed, different and greater, pressures for a bridge. By 1918, city planners were considering spans across San Francisco bay as an alternative to the existing ferry system. A system of antiquated ferries was operated by the Southern Pacific Railroad without competition and was little more than a collection of boats jerry-rigged to carry automobiles. The vessels crossed the bay slowly, infrequently, and charged high tolls.

Beginning in 1921, several independent ferry companies took advantage of high demand for travel across the bay by providing fast and frequent service on diesel powered ships specifically designed to carry cars. This new service, with reduced tolls, quickly gained market share, and customer demand steadily increased along with new capacity. Attempts by the Southern Pacific Railroad to buy out these new operators created further interest in automobile travel across the bay. In May 1929, Southern Pacific briefly regained control of all cross-bay ferry service by purchasing all of the operators and merging them together. Travel on the ferry system had increased by over 700% in one year.

The unification of the ferry system allowed Southern Pacific to better match capacity to demand, further improving overall system capacity. Even so, by the time of the merger, demand had again expanded beyond the new capacity. Sunday night traffic jams on the highways of Marin County leading to the Sausalito ferry terminal attested to the pent-up demand for travel across the Golden Gate. As Southern Pacific continued its efforts to control and expand the ferry business, the municipal government of San Francisco passed local ordinances that would make a bridge across the mouth of the San Francisco Bay – the Golden Gate – possible.

Location of the Bridge

Geographically, the mountains of Northern California divide the state into long north-south valleys and intermittent high and low passes. The action of the rivers, now called the Sacramento and the San Joaquin, carved out the pass that now forms the Golden Gate. This allowing the Pacific Ocean to flood the 463 square miles of basin, leaving the peninsula of present day San Francisco surrounded on three sides by water (See Exhibit 1-1 on the companion CD). The valley that ran north-south though Marin County to San Francisco provided a natural crossing at the Golden Gate. A bridge crossing at such a location forms a natural geographic monopoly to access San Francisco from the north.

Exhibit 1-1   Map of San Francisco Bay Area

Exhibit 1-1 By Permission, Golden Gate Bridge and Highway District, from *The Golden Gate Bridge*, Report of the Chief Engineer, 50th Anniversary Edition, 1987.

Assessing the Bridge's Feasibility

Studies for the Golden Gate Bridge began with a 1918 City Council resolution ordering the feasibility study of a bridge across the Golden Gate. The city engineer, M. M. O'Shaughnessy, approached Joseph B. Strauss, a nationally recognized bridge engineer to look at the project. O'Shaughnessy was not convinced that the project was technically or economically viable.

"Everybody says it can't be done and that it would cost over $100,000,000.00 if it could be done."[1]

Strauss disagreed and thought that a suitable bridge could be constructed for a more affordable price after carefully studying site conditions at the Golden Gate.

Following a study jointly undertaken by the City and the United States Coast & Geodetic Survey,[2] a Request for Proposals (RFP) was sent out in May 1920 from O'Shaughnessy's office soliciting designers and cost estimates for the project. The RFP was sent to Joseph Strauss in Chicago, Francis C. McMath of the Canadian Bridge and Iron Company of Detroit, and Gustav Lindenthal, the engineer for the 1000-foot Hell's Gate Bridge over the East River in New York in 1916.[3]

McMath never officially responded. Lindenthal estimated that the bridge would cost between $60 and $77 million dollars. Strauss, however, estimated a total cost of $27 million. Strauss' estimate, although over the $25 million budget proposed by the city, brought the discussion from theoretical possibility to serious consideration.[4] Strauss was selected as the Project Engineer.

Among the many obstacles facing the bridge at that point, the most important were the rail/ferry monopoly and the military, which owned the land at each end of the proposed bridge. Southern Pacific was then in the midst of its effort to regain its share of the ferry traffic from the independent operators. Southern Pacific was investing heavily in improved ferry services and was loath to let a fixed link siphon off all ferry customers. Exhibit 1-1, on the companion CD, shows the results of a 1935 traffic study which projected traffic flows that might use the new bridge.

Exhibit 1-2 By Permission, Golden Gate Bridge and Highway District, from *The Golden Gate Bridge*, Report of the Chief Engineer, 50th Anniversary Edition, 1987.

The Army and Navy had different concerns. First, the approaches to the bridge would, by design, pass through the Army's Presidio base. The Navy wanted to insure that its major naval base in San Francisco would continue to have access to the Pacific Ocean.

Because the use of military land was crucial to the success of the project, one of the first steps taken in the planning process was to apply to the Secretary of War for a permit to build the bridge and make use of federal land needed for the bridge structure and approach roads. The application was filed by the "Bridging the Golden Gate Association" and representatives from San Francisco and Marin Counties before the group had official authority to build the bridge. Colonel Herbert Deakyne held a hearing on the petition in May 1924.[5] Based on Deakyne's favorable report, the Secretary of War, Weeks, issued a provisional permit in December 1924 to proceed with the bridge, pending submission of detailed plans.[6]

The Formation of the Golden Gate Bridge Highway and Transportation Authority

The Bridge and Highway District Act created the Golden Gate Bridge Highway and Transportation Authority (the "Authority"), authorized multiple counties on both sides of the Golden Gate to vote to allow the District to borrow money secured by taxes on private property in the supporting counties. The ability to tax would defray the initial expenses and provide lending institutions with valuable assets to secure debt repayment if the toll revenue failed to meet the bond payment obligations. The Act was written so that citizens of the counties within the District could, by vote of each county, include or exclude each county in the District. (See Exhibits 1-4 through 1-6 on the companion CD for information on the bridge bonds)

Exhibit 1-3   Summary of the Bridge and Highway District Act of California

After the 1924 Act was passed, forces in opposition to the bridge brought suit in every county listed in the District: San Francisco, Marin, Sonoma, Napa, Humboldt, Mendocino, and Del Norte. The litigation continued for nearly six years, during which time Humboldt County and portions of Mendocino County dropped out as members of the District. The factions opposed to the bridge argued that it would be far more expensive than Strauss' $27 million estimate and that the tolls would never be sufficient to pay for interest and principal on the bonds sold to finance the initial construction and for maintenance of the structure. The claim was that taxpayers in the District would have to pick up the difference; i.e. that individual property owner in the various counties would be taxed to retire the debt. Another argument against the bridge was that it would slow future development of San Francisco by limiting the height of ships entering the harbor to 200-feet.

On December 4, 1928, after the first round of litigation was concluded in favor of the District, the District was formally incorporated under the 1924 Act and the process of getting approval from member counties to issue the bonds needed to fund the bridge began.

Bidding the Project

The District's first action, in July 1929, was to increase the tax rate of 3¢ for every $100 in taxable property in the District. The purpose of the tax was to fund planning and early design of the project. The tax fueled new opposition arguments. "Taxpayers forced to outrageous payments for the bridge." The arguments were forcefully made by experts and men of standing in the community. The District responded by pledging that the entire project would not cost more than $35,000,000. Just months later, in November 1930, votes in the District approved the first bond issue to finance the bridge by a vote of 145,057 to 46,954.[7]

To increase the level of confidence that the entire project could be built in the proposed budget, Strauss obtained firm fixed quotes for all major components of the project. This practice was unusual in 1930, but commonplace today.

"Pursuant to their pledge, therefore, the Directors ordered the Chief Engineer (Mr. Strauss) to prepare plans and specifications covering all units of the work so that lump sum bids could be received and the construction cost of the project ascertained in advance of the awarding of any contracts. Such procedure, of course, is contrary to standard practice on major jobs, where the starting of certain units of construction must, in the nature of things, be postponed until other units have been finished."[8]

Strauss included an early version of the "Changes" clause that is now standard in public construction contracts throughout the United States. He went on to say,

"[We] realized also that proper provisions should be made in the proposals and contracts, whereby adjustments in the contractors' compensation because of changes in plans could be on a basis agreed upon in the contracts..."[9]

The contractors examined Strauss' Invitations for Bids in one of the most trying times of United States history. By the summer of 1931, the depression weighed heavily over America. Things were clearly going to get worse before they got better. The bids, returned in July 1931, were as shown in Table 1-1.[10]

I-A	Steel Superstructure	$10,494,000.00
I-B	Steel Cables, Suspenders & Accessories	$6,255,767.65
II	San Francisco Pier & Fender and Marin Pier	$2,260,000.00
III	Anchorages & Piers of Approach Spans	$1,645,841.28
IV	Steel for S.F. and Marin Approaches	$996,000.00
V	Presidio Approach Road	$966,180.00
VI	Sausalito Approach Road	$67,586.00
VII	Paving of Main Span, Side & Approach Spans	$345,000.00
VIII	Electrical Work	$133,495.00
IX	Toll Houses & Service Buildings	$71,430.00
X	Cement (Est. 500,000 bbls.)	$1,220,000.00
	Total	**$24,455,299.93**

Table 1-1   Bids Returned, July 1931

The bids indicated that if Strauss' design documents remained stable, the costs of construction alone would come in under not only the bond issue, but also under Mr. Strauss' estimate of $27 million. When the engineering fees, legal fees, and interest were included, total project costs were much closer to $35 million.

Bond Financing for the Project

The six-year fight over the legality of the District proved only to be a prelude to the opposition's efforts to stop the bridge. As outlined in the Bridge and Highway District Act, the District had the right to issue bonds, subject to approval by vote in the represented counties. Through numerous articles and advertisements in the local newspapers, the opposition attempted to dissuade the public from supporting the bond issue. The arguments included those used in the legal fights against the formation of the District. The press was sprinkled with new insinuations that the Directors of the District and the Chief Engineer (Strauss) were going to profit personally from the project.[11] (See Exhibits 1-4 through 1-6 on companion CD for information on the bridge bonds)

Exhibit 1-4   *Newspaper Advertisement on the Bridge Bond Issue*

Exhibit 1-5   *Bond Sales Record*

Exhibit 1-5 By Permission, Golden Gate Bridge and Highway District, from *The Golden Gate Bridge*, Report of the Chief Engineer, 50th Anniversary Edition, 1987.

Exhibit 1-6   *Bond Payment Schedule*

Exhibit 1-6 By Permission, Golden Gate Bridge and Highway District, from *The Golden Gate Bridge*, Report of the Chief Engineer, 50th Anniversary Edition, 1987.

After the District obtained voter approval for issuing the bonds, one of the contractors and a group of proposed investors raised questions regarding the constitutionality of the District's right to tax. These questions alarmed the financial community since the creditworthiness of the District and its bonds depended on the security pledged by taxpayers in the member counties. To solve this new problem, the Supreme Court of California was asked to decide this question before the bonds were issued and the contracts let. The Justices, noting the urgency of the matter, heard the case quite quickly and, with one Justice dissenting, ruled for the District.

Still, the legal struggle was not over. Yet another suit, by two companies from the northern parts of California, was brought against the District, challenging the right of the District to tax for the purpose of servicing the repayment of principal and interest on the bonds. Garland Co. v. Filmer, et al. (1932 N.D.Cal.) 1 F.Supp 8. The District won the case at the trial level and an appeal was withdrawn in 1932.

This seemed to be the last legal hurdle left to issue the bonds and construct the bridge. But another difficulty arose as steps were begun to let the contracts and to begin construction. Bankamerica Company, the lead in a syndicate organized to finance the bridge, bid 92.3 cents on the dollar for the 4.75% bonds that were to be sold by the District to finance the bridge. The District's authorizing statute allowed only a 5% interest rate on the bonds. Even though the bonds were issued at 4.75%, Bankamerica's offer at 92.3 cents pushed the effective yield to 5.25%. Bond attorneys from New York, retained by the District, thought that the Bankamerica offer, if accepted, would make the bonds illegal. Counsel local to San Francisco disagreed, but the District decided that the issue required attention. Unfortunately there were no funds left for the counsel to bring the matter before the appropriate court for a decision.

Strauss approached the chairman of the Bank of America, A. P. Giannini and, with the help of the most powerful people associated with the bridge's interests, convinced Giannini to form a new syndicate for the purpose of buying Golden Gate Bridge Bonds. The syndicate was formed and bravely, considering the depression bond market, agreed to buy $3,000,000 worth of bonds from the District at 96.23 cents on the dollar, enough to put the effective interest rate just under 5%. The newly formed syndicate agreed to forward $184,600 against $200,000 worth of the $3,000,000, pending the outcome of the legal question regarding the effective percentage rate. Funding for the preliminary planning and early design was now available and work continued. This last difficulty was resolved and the District was finally in a position to proceed with construction.

Construction

Construction of the bridge involved hundreds of contractors and subcontractors and many bridge-specific castings and forgings. There were many opportunities for the potential problems raised by the bridge opponents to come true and for expenses to run out of control. Despite several delays and cost overruns, the bridge remained within the first estimate. Some costs were also saved when part of the San Francisco approach was undertaken as a city project. Responsibility for that approach was shifted when the city objected to the planned connection at Marina Boulevard and insisted that the road connect with Lombard Street. In addition, responsibility for a portion of the approach on the Marin side was shifted to the state government. Both of these changes helped The District keep their costs for the bridge under the $35 million ceiling.

In many cases, construction techniques needed to be altered or invented anew to meet the needs of this bridge. The San Francisco-side foundation pier, built essentially in the open ocean, required an access trestle 1100 feet long and 22 feet wide running out into the mouth of the Golden Gate. The trestle was rammed by an off course vessel, tearing out a section of the structure and seriously damaging the ship. The trestle, seriously weakened by the encounter with the ship, was further damaged in a severe storm and lost over 2/3 of its length into the ocean. Rebuilt, the trestle remained for the duration of construction, allowing access to the San Francisco pier via a newly secured sub-structure tied to the bedrock with steel cables.

Plans for the San Francisco pier itself underwent four expensive revisions before a final construction approach was chosen. The original plan was to pour a concrete fender that would protect an oval section of water out in the bay. Within that fender, a pneumatic caisson would be used to reach dry ground at the bottom of the bay. Construction would then progress as if the tower pier were on dry land rather than 1,200 feet out into the ocean. However, when the fender was partially complete, and the caisson was floated in from the open eastern half, the wave action inside the half-enclosed fender was sufficient to nearly destroy the caisson. A quick decision was made to abandon that method of construction and the caisson was floated back out without undue damage to the fender. The entire custom-built pneumatic caisson was later sold for scrap.

The San Francisco pier eventually cost as much as the estimate for both piers, and that was but one cost overrun. The repairs and adjustments to the military reservations, originally estimated at $100,000, eventually ran to $575,000. Military cooperation was crucial to the project. Continued approval was required for the access roads, approaches, and tollhouse area. Every new application for changes in use was met with additional complications. The military requisitioning procedure meant massive delays and cost overruns. By the end of construction, The District had paid for the removal and rebuilding of barracks, roads, sewage and drainage systems, fire stations, gas stations, a new powder magazine (costing $125,000.00), machine shops, and a new rifle range.[12]

The cable saddles that guide the cables (in an octagonal formation) over the tops of the towers, the cable bands that clamp down on the main cables and attach the suspender ropes, and the strand shoes that allow the main cable strands to attach to the anchorages were all custom castings. Each one was cast and machined to a smooth finish, allowing even bearing between the cable and casting. These were all subject to testing for cracks and voids. The cable bands in particular were notorious for voiding (hardening with air pockets deep within the metal) and had to be re-cast many times over. The errors in casting were only evident when the part was finished in the machine shop, but at that point the expensive work had already been done. The cable bands, and the rest of the castings were very expensive.

During construction, battles of public opinion continued to be fought in the local papers. Once, during the construction of the San Francisco pier, a local geologist questioned the seven experts' opinion regarding the safety of the founding of the pier. The directors of The District were brought by the tide of public fear to a public forum where they refuted the claims of the geologist. Early on, the Chief Engineer, Joseph Strauss, formed an information campaign that remained active from before the formation of The District until the bridge was completed.

Exhibit 1-7   By Permission, Golden Gate Bridge and Highway District, from *The Golden Gate Bridge*, Report of the Chief Engineer, 50th Anniversary Edition, 1987.

Questions

For simplicity, ignore the effects of inflation in the following questions.

1.   Compute Your Own Estimate of Net Present Value

Familiarize yourself with Exhibit 1-8 on the companion CD.   This worksheet contains the initial assumptions relating to cash flow for the project <u>prior</u> to the letting of contracts.

Exhibit 1-8   Calculations Worksheet

Copy the worksheet into your own file and fill in the balance of the cash flows.   State your assumptions regarding the timing of the bond sales, the distribution of construction costs, and any other assumptions you make in filling out the worksheet.

The worksheet includes reduced toll rates over time.   This was included in the project cost and revenue estimate.   Exhibit 1-9 shows spikes in toll receipts over time (the top curve).   If rates are going down and revenues are reasonably flat, what does this tell you about the projected traffic volume?

Calculate the NPV of the cash flow you generate.

Exhibit 1-9   Forecast of bridge traffic and revenue
Exhibit 1-9 By Permission, Golden Gate Bridge and Highway District, from *The Golden Gate Bridge*, Report of the Chief Engineer, 50th Anniversary Edition, 1987.

2.   Sensitivity to Changes in Assumptions

Key assumptions made by Strauss regarding toll pricing eventually proved to be wrong.   You are not surprised, are you?   For example, in response to the completion of the Bridge, in 1936, the ferries dropped their prices.   The Golden Gate Bridge Authority was forced to adopt a lower toll rate than assumed in the forecasts.

Assumptions used in any financial forecast are just that. Question 2 tests the sensitivity of the NPV you calculated in Question 1.

Compute the NPV of the cash flow for each of the following cases, and compare it to the base case you calculated in Question 1.

a. A 10% increase in toll paying vehicles across the entire cash flow.

b. A 10% decrease in toll paying vehicles across the entire cash flow.

c. A delay in construction completion of 2 years (at the same total cost).

d. An increase in the total construction price of 10% (allotted in the same proportions as your base case from Question 1).

e. A 25% increase in the planned toll rates.

f. A 50% increase in the planned toll rates.

3. Answer the Following Questions.

a. What was the project delivery method for the Golden Gate Bridge? Did the Counties "outsource" the project's cash flow (sources and uses) to an independent entity? Why? Is the Authority a public entity? A private entity?

b. In which quadrant does the Golden Gate procurement fit?

c. Some components of the project were not completed within the original estimate. For example, the San Francisco pier experienced a 25% overrun. The military refurbishments were nearly 500% over budget. What degree of cost overrun for the whole project would have caused the "failure" of the project? (How much cost overrun would cause the project to "fail"?)

d. Put yourself in the shoes of Bank of America. Why would you agree to buy the bonds at such a rate? What were your Risks? Potential Rewards? Does your NPV analysis help justify these risks?

e. Based on the NPV results, did citizens have to worry about the Authority's ability to repay the bonds? Were the tolls capable of handling debt service payments?

References

Cassady, S. Spanning the Gate. Mill Valley, California: Squarebooks, Inc., 1979.

Golden Gate Bridge and Highway District. Annual Reports 1937/38 – 1979/80. San Francisco, CA: Golden Gate Bridge and Highway District, 1938-1980.

Newspaper Scrap-Book. (Articles principally from the San Francisco Examiner and the Santa Rosa Independent). Property of The Golden Gate Bridge and Highway District, San Francisco, CA, ca. 1928 – 1937.

Strauss, J. and C. Paine. The Golden Gate Bridge; Report of the Chief Engineer to the Board of Directors of the Golden Gate Bridge and Highway District, California. San Francisco, CA: Golden Gate Bridge and Highway District, 1938.

Notes

[1] Report of the Chief Engineer, J. B. Strauss, p. 25

[2] Report of the Chief Engineer, J. B. Strauss, p. 27

[3] Spanning the Gate, S. Cassady, p. 43

[4] Report of the Chief Engineer, J. B. Strauss, p. 27

[5] The group that applied to Colonel Deakyne did so based on the supposition that an official district would be formed to administer and finance the bridge. That district was made possible by the Bridge and Highway District Act of California, May 25, 1923.

[6] That provisional permit eventually saved the project in 1930, when final approval was sought before General Lytle Brown, who opposed the bridge but could not rescind the provisional permit presumably because to do so would dishonor General Deakyne's judgment (Report of the Chief Engineer, J. B. Strauss, pp. 31 & 39). It is difficult to imagine that the same line of thinking would be allowed to prevail in today's environment.

[7] Report of the Chief Engineer, J. B. Strauss, p. 42.

[8] This illustrates an early of the Design-Bid-Build concept. Clearly Mr. Strauss identifies standard practice as evolutionary, where units are bid in sequence, as they become needed. This probably allowed for a greater degree of flexibility in design and construction, but greater uncertainty in cost.

[9] Report of the Chief Engineer, J.B. Strauss, p. 44. Note that this is an early use of a changes clause on a major infrastructure project.

[10] Report of the Chief Engineer, J. B. Strauss, p. 44

[11] Most signature projects like the Golden Gate Bridge produce similar allegations-some real, some imagined. The Brooklyn Bridge, the New York Subway, the Eads Bridge, and the Central Artery Project in Boston are examples.

[12] Report of the Chief Engineer, J. B. Strauss, p. 51

CHAPTER 2   THE DULLES GREENWAY

Infrastructure Development Systems IDS-97-T-014

Professor John B. Miller prepared this case with the assistance of Research Assistant Om P. Agarwal as the basis for class discussion, and not to illustrate either effective or ineffective handling of infrastructure development related issues. The assistance of Michael R. Crane of TRIP II and Frank J. Baltz, Esq. of Shaw Pittman & Potts in the preparation of this case is gratefully acknowledged. Thanks to Charles Y.J. Cheah for adding the Monte Carlo materials. Data presented in the case has been altered to simplify and focus the issues, and to protect the confidentiality of those who were kind enough to help in describing the project.

Introduction

John Smith, Chief of Financial Services for CIGNA Investments was in the middle of a tough week. The Dulles Greenway project (the first private toll road to be designed, built, and operated in the United States in over a century) was in trouble. Difficult decisions lay ahead for the consortium of banks Smith had put together to finance the project, the equity contributors to the project, the developers, and the operators. Forecasted traffic, thought by the Banks to indicate a solid revenue base to support operations, maintenance, and debt service, had not yet materialized. Revenues were a fraction of what had been projected in the first two years of operation. Reductions in the tolls had produced more users each day, but, unfortunately, had not dramatically increased net revenue. For the first eighteen (18) months of operation, John had remained hopeful that traffic levels would grow, at least to the point where operations and maintenance costs were covered, and bank debt could begin to be serviced out of operating revenues. However, in early 1997, word had come from the Virginia Department of Transportation (VDOT) that it would proceed with the expansion of Route 7, a nearby competing public road.

Several of John's colleagues from the consortium of funding banks had called. The questions they raised were not very different from those on John's mind. None of the short-term alternatives seemed very attractive. What was CIGNA planning to do if debt were not serviced in short order? Should the banks restructure the debt? Foreclose on the loans? Take over the project? Might some of the financing banks sell their position to others, and if so, with or without a discount? Did the project continue to have potential in the long term, especially if real estate development continued in the area between Dulles International Airport and Leesburg? The franchise agreement with the state allowed profits of between 14% and 15% above debt service, and, in the event lower profits were earned early in the franchise, such shortfall could be made up later in the franchise. The continuing success of the Broadlands project, a local real estate development of 4000 dwelling units and 500,000 square feet of retail space at one of the Greenway exits, could produce a substantial long term improvement in the performance of the Greenway. On May 17, 1997, a consortium of EPC firms proposed to extend/connect the Washington Metro from the West Falls Church station to and through Dulles Airport to a new Ashburn station in the median of the Greenway. Should the banks stick with the project for the long haul? Sell part of this long-term risk at a discount? Get out entirely? Was the decision to fund this project sensible in the first place? John retrieved his files on the project, including his closing binder from the financial transaction, and many news articles that had appeared in the local press before, during, and after delivery of the facility.

Background of the Project

The Dulles Greenway is an extension of the existing Dulles Toll Road from the entrance to Dulles International Airport west for approximately 14 miles to the junction of Routes 7 and 15 near Leesburg, Virginia. The current Dulles Toll Road runs alongside the Dulles Airport Access Road built by the FAA in the early 1960's at the same time as the airport. Both the Dulles Toll Road and the Dulles Airport Access Road connect the area around the Airport with Tyson's Corner, Interstate Route 495 around Washington, and Interstate Route 66 to Washington D.C. The Airport Access Road is free, but is restricted to vehicles visiting the airport. The

Dulles Toll Road was constructed in the mid 1980's to relieve congestion caused by suburban commuters residing in Loudoun County, Virginia, who worked in the metropolitan Washington area. Figure 2-1 shows the area surrounding the project.

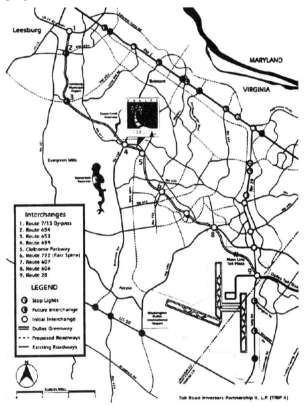

Figure 2-1   Map of Area Surrounding the Dulles Greenway

Reprinted by Permission of TRIP II

The Dulles Toll Road Extension (which became the Dulles Greenway) was originally conceived to be a critical link in the transportation network between Fairfax and Loudoun counties. State Routes 7 and 50 were the only major east-west arterials connecting the growing population of Loudoun county with rapidly growing employers at Tyson's Corner, and in Fairfax, Arlington and Washington, DC. Route 28 provides North, South movement between Routes 7 and 50, but was also the principal route for access to Dulles International Airport from the west. For a typical Loudoun county commuter, the trip from Leesburg to Tyson's corner included 14 traffic lights before the Dulles Toll Road was reached. The Toll Road itself, which included frequent stops at toll booths for the payment of small increments of the toll, was extremely congested throughout both rush hours, making for a difficult, time consuming, frustrating daily commute.

In early 1987, the Virginia Department of Transportation (VDOT) embarked on a series of studies, hearings, and citizen information meetings to establish an acceptable alignment and to conduct appropriate environmental analyses for constructing a four-lane, limited access roadway from the intersection of Route 28 and the Dulles Airport Access Road northward to the junctions of Routes 7 and 15 in Leesburg, Virginia. Throughout these studies, the proposed project was known as the Dulles Toll Road Extension.

As a result of a Commission established by the Governor of Virginia in 1987 to address a $7 billion shortfall in public funding for needed transportation projects, in 1988, the Virginia legislature passed a state authorizing a private corporation to build, own, and operate a toll road for profit. The statute requires that the State Corporation Commission, not the Department of Transportation, make a determination that approval of any private application to construct such a facility be in the public interest, and that the Commonwealth Transportation Board (VDOT) approve the project's location, design, and construction costs. Later that same year the Commonwealth Transportation Board approved the Route 28 to Routes 7 and 15 proposed alignment of the Dulles Toll Road Extension. In the spring and summer of 1989, the Toll Road Corporation of Virginia (the predecessor firm to the current owners of the Dulles Greenway) submitted an application to build the project. Hearings were held in both Loudoun and Fairfax counties in May and June 1989, and the application of TRCV was approved by the State Corporation Commission in July 1989. The Commission later issued a certificate of authority in June 1990. Between 1990 and September 1993, the certificate of authority was transferred to a successor entity, Toll Road Investors Partnership II (TRIP II), conceptual designs were submitted and approved, and financing was arranged. Environmental approvals were sought and obtained. A resolution was adopted by the State Corporation Commission approving a construction cost of $293.8 million, and a construction schedule commencing not later than September 30, 1993 with completion not later than March 31, 1996.

The name of the project was changed to the Dulles Greenway. The project is a four lane, limited access highway located within 250 feet right of way. The project includes seven interchanges along the route, with two more scheduled to be added in the future, when traffic volumes reach 93,000 and 132,400 vehicles per day. The facility was designed to permit lane expansion, and even contemplates a future extension of mass transit in its median strip. State of the Art toll collection technology was incorporated into the design to permit toll collection through windscreen transponders without drivers having to slow down or stop. The developers of the project, which is privately owned and financed, assembled the required parcels of land through purchase, donations, and, in the case of the Metropolitan Washington Airports Authority, through lease. See Exhibit 2-1 for a summary of the key events leading to the development and operation of the Dulles Greenway.

Exhibit 2-1   Dulles Greenway Milestones

Financing the Project

Most of the financing for the project came through a consortium of ten institutional investors led by three major project investors: John's company (CIGNA Investments Inc.), Prudential Power Funding Associates, a unit of the Prudential Insurance Company of America, and John Hancock Mutual Life Insurance Company. The consortium provided $258 million of long term fixed rate notes, due 2022 and 2026, to finance much of the cost of construction and initial operation of project. A second group of banks, consisting of Barclay's Bank, Nation's Bank and the Deutsche Bank provided a portion of the construction financing and a $40 million revolving credit facility.

The Owner/ Developer group (TRIP II) was a partnership comprised of three entities, the Shenandoah Greenway Corporation, Autostrade International, S.p.A., and Brown & Root. The Bryant family controls

Shenandoah, contributing approximately $22 million in equity into pre-construction project development between 1989 and 1993. Autostrade is an American subsidiary of a French corporation specializing in the operation of European toll roads. Brown & Root is the constructor of the project. Autostrade and Brown & Root contributed approximately $16 million in equity to the project. TRIP II also arranged to provide up to $46 million in irrevocable lines of credit to the project, in the event that revenues were insufficient to (a) operate and maintain the facility, and (b) pay debt service on both the construction and permanent debt supporting the project.

The debt financing was secured by a first mortgage and security interest in the TRIP II's entire right, title and interest in the project. Construction lending and the revolving credit facility were to be repaid first out of the net project revenues, after unalterable payments of maintenance and operations costs and fixed lease charges. Next in repayment priority was the institutional lenders' long term fixed rates notes, followed by repayment of advances, if any, from TRIP II lines of credit. Only then would remaining net revenues be available to pay returns on equity.

The project was estimated to cost approximately $326 million, of which $258 million represented the cost of design, construction, and toll collection equipment to put the project in service. The balance represented pre-construction development costs and interest during construction.

Prior to closing the loan, John made his own personal rough calculation of how the facility would perform financially John's calculations indicated that the project would achieve a Debt Coverage Ratio of 1.5 within 6 years of operation. John calculates the Debt Coverage Ratio by dividing Operating Revenues in a given year by the sum of principal and interest payments on outstanding debt for the same year. Ratios of 1.3 to 1.5 are normally considered "good" in his experience. Key assumptions made were these:

1. First year total of 34,000 fares collected per day at $2.00, with 10% traffic growth per year in years 2 through 13, up to 93,000 fares/day in year 14. A 4% per year growth in traffic in years 14 through 22, with a maximum capacity at 132,400 in years 23 through 43.

2. An 8% average discount rate.

3. An annual fare escalation in Years 4 through 43 at a rate equivalent (over the term) to 2.5% per year.

4. O&M expenses (including, without limitation, resurfacing, police enforcement, and toll collection) of $7,000,000 in year one, escalating at 5% per year.

5. A Fixed Annual Lease Payment by TRIP II of $500,000 to the Metropolitan Washington Airport Authority, beginning in the year the road opens.

6. Phase 2 and Phase 3 Improvements to the road of $20 million and $40 million respectively in years 13 and 23.

Figure 2-2 shows the financial structure for the project, in terms of how gross toll revenues are distributed to the project's upkeep and to various stakeholders. An O&M irrevocable reserve, part of "Annual Costs," cannot be diverted. Thus, O&M is the first commitment of all stakeholders, since it preserves and protects the facility throughout the period of debt service and helps to ensure the condition of the road (for the State of Virginia).

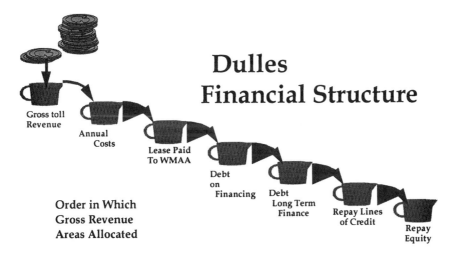

Figure 2-2   Financial Structure for the Dulles Project

The approvals for the project cap the investors' aggregate rate of return at approximately 18% until the initial debt is paid. With that aggregation, the rate of return on investment varies from 8% to 30% for the various forms of debt, equity, and irrevocable lines of credit. With these assumptions, the parties involved had concluded that the project was viable.

Delivery of the Dulles Greenway

Project Delivery Method and Execution. The project delivery team had extraordinary success in producing the project: ahead of time, within budget, with high quality, and with early recognition and award-winning treatment of environmental issues. Though scheduled to take 30 months for completion, the road had been delivered 6 months ahead of schedule, in September 1995. Although there were some disagreements among the project team during design and construction, John could not imagine a stronger record of performance. Once the project's approvals and financing package was in place, the project was delivered significantly faster than a typical public interstate highway. The Dulles Toll Road Extension was going to be built, if not in the early 1990's by TRIP II, then at some future time by the state using tax dollars. John remained convinced that the delivery process employed on the Greenway, a forty-two and one half (42.5) year Design-Build-Finance-Operate franchise awarded by the state, was well worth emulating on appropriate public infrastructure facilities throughout the nation. The trick, however, was in identifying those appropriate circumstances.

Pre-Construction. On the financing side, the "handshakes" between the Developer and the State, and the Developer and the Financing Banks, seemed in retrospect, to be cumbersome, confusing, and difficult. The project was delivered by the Developer through competitive processes, which John felt confident had produced a high quality project in a short time for a fair price. The irony was that an extraordinary success on the delivery of a publicly desirable expressway might turn out to be inappropriately financed in the private sector.

John could not help but wonder whether streamlining the approval and financing process would have materially changed how the project was performing now. Had the new state statute governing approval of the project been a help or a hindrance to a streamlined delivery of the project? Did the sole source nature of the developer's proposal unnecessarily drive up development costs, because there were no competitive proposals which could independently confirm to the state and to financing banks the viability of the project's expected price, quality, and schedule? Prior to the start of construction, these risks had seemed large, consuming a great deal of energy, time, and resources for which the developer's paid. In hindsight, the design/build team had proven to be extraordinarily capable of managing these risks. Had several false starts on the financing been avoidable, or were they the natural result of the new procurement process?

Pre-1990 Ridership Forecasts. A variety of traffic forecasts were done by independent consultants prior to the commencement of construction, based upon a 1989 start of construction and facility opening in 1992. In part these forecasts were based upon the economic boom of the 1980's. These forecasts did not anticipate the dip in the economy of 1991, nor did they anticipate that four years would be required to (a) obtain approvals from VDOT, the State Corporation Commission, cognizant state and federal permitting and environmental authorities, (b) complete an appropriate preliminary design for approval purposes, and (c) assemble financing. In 1989, consultants were projecting that ridership would be 20,000 cars per day in the first year of operation (1992) at a toll of $1.50. By the fourth year of operation, 1995, ridership was projected to be 34,000 per day at the same toll rate.

Operating Results

The facility opened in September 1995, six months ahead of schedule. The project was completed in two years. The initial, introductory fare was $1.75 for the last four months of 1995. The toll rate was scheduled to be raised to $2.00 as of January 1996. By the end December 1995, paid traffic on the facility was approximately 10,500 per day, with a growth rate of approximately 1% per month (12% per year). The decision was made not to raise the fare to $2.00 in January 1996, and in fact, the fare was lowered to

$1.00 in March 1996. Ridership grew to 21,000 per day as of July 1996, with a steady increase of paid users of approximately 1% per month.

Gross toll revenues have not been sufficient to make debt service. The parties are attempting to restructure the project on mutually acceptable terms. As recently as March 29, 1997, the Washington Post reported that the project's developers and creditors were close to an agreement for restructuring the debt on the project, in part through a moratorium on the payment of debt service, and in part through an $0.15 increase in the current $1.00 toll rate.

Questions

1. **1995 Cash Flow Expectations.** Become familiar with the spreadsheet labeled *Cash Flow Projected Sept 93*, Exhibit 2-2. This spreadsheet includes a discounted cash flow analysis of the Dulles project. The spreadsheet allows its user to alter the cash flow projections by adjusting variables in the upper left corner of the worksheet.

Exhibit 2-2   Cash Flow Projected, September 1993

   Using Exhibit 2-3, labeled *Cash Flow Actual Dec 95,* adjust the worksheet so that it reflects the actual data <u>as of December 1995.</u>

Exhibit 2-3   Cash Flow Projected, December 1995

   When will the project be able to make payments on debt service? Should the Improvements in years 13 and 23 be undertaken? If so, when? What is the expected NPV for the project? What is the debt coverage ratio for the project? Make reasonable assumptions and state them separately or show them in the worksheet.

2. **1996 Cash Flow Expectations.** In the Dulles workbook, there is a third worksheet *labeled Cash Flow Actual March 1996,* Exhibit 2-4.

Exhibit 2-4   Cash Flow Actual, March 1996

Using this worksheet, adjust the worksheet so that it reflects the actual data as of March 1996.

The project developers were required to put up a line of credit of $40 million as additional callable "equity" in the event revenues were insufficient to service the debt. How much, if any, of this line of credit is likely to have been tapped as of March 1996. Should the Improvements in years 13 and 23 be undertaken? If so, when? What is the expected NPV for the project? What is the debt coverage ratio for the project? Make reasonable assumptions and state them separately or show them in the worksheet.

3. **Debt Capacity as a Function of Interest Rate with Static Net Revenue.** Assume that the original Cash Flow Projected in September 1993 of $17,320,000 in Net Revenues after M&O and Lease Payments are made is correct. Prepare a plot of Debt Capacity on the y-axis v. Interest Rate (ranging from 1% to 25%) on the x-axis based on the assumption that $17,320,000 is available each year for 30 years to service the initial capital debt. In other words, determine how much you can borrow against the annuity of $17,320,000 at varying rates of interest and graph the results. (EXCEL has functions that will do this for you.)

4. **Sensitivity.** Use Exhibit 2-5, *Sensitivity Work*, to explore how sensitive the project NPV is to changes in:

    a. the assumed discount rate?. Plot changes in NPV for discount rates varying between 4% and 20%.

    b. the initial toll charged? Plot changes in NPV for initial tolls varying in $.25 increments between $.50 and $2.00 (assuming the same 2.5% increase per year shown in the existing spreadsheet.

    c. the number of toll paying vehicles? Plot changes in NPV for initial traffic volumes (in 5,000 increments) ranging from 5,000 per day to 40,000 per day.

    d. the cost of construction? Plot changes in NPV for initial capital costs (evenly spread over years 1 and 2) ranging from $250 million to $400 million. [not larger than 25 million increments].

Exhibit 2-5   *Sensitivity Work*

Summarize in one or two well-organized paragraphs your conclusions about the sensitivity of the project to changes in discount rate, tolls, vehicles, and initial cost.

5. **The Effect of Extraordinary Transaction Costs on Project Viability.** Assume $38 million dollars was invested by the developers over the four year period between the date the initial application to build the project was filed and the date construction started. Such costs are typically described as "Transaction Costs", i.e. costs associated with the transaction, but not associated with actual design, construction, or operations.

What if $30 million of the $38 million shown in the first worksheet *Cash Flow Projected Sept 93* could be used to help cover a portion of the $258 million capital costs of the project, and that the project was constructed in the 1989 to 1991 period as originally planned. What is the NPV of the project if the 1989 traffic forecasts and initial toll rates are used?

6. **Procurement Analysis.** Where does this procurement fit in the Quadrant framework? Did the Commonwealth conduct a competitive process for the award of this franchise? Would competitive proposals have helped the Commonwealth? TRIP II? Other potential proposers? Why or why not? How? Is it in the "public interest" of the Commonwealth of Virginia to award a franchise to build and operate this type of facility without an independent verification (through competition or otherwise) of project viability? In light of the extraordinary expenditures of transaction costs on this project, who should control scope – the owner or potential developers?

7. **Conclusions.** What would you recommend that John Smith and CIGNA do now? Could CIGNA conduct a procurement to better

consider its options? What approach do you suggest? What type of analysis do you recommend?

8. **Monte Carlo Simulation** (For illustration only)

This task dwells deeper into the topic of risk assessment and its impact on financial decision making.

In Question 4, a sensitivity analysis of the project, subjected to changes in discount rates, tolls, vehicles and initial cost, has been introduced. It is important to realize that sensitivity analysis only provides a "static and deterministic" view of how various factors can affect the outcome of a project. In reality, there is some randomness associated with each of these factors. Monte Carlo simulation is one approach that allows the analyst to incorporate the stochastic behavior of these factors into the analysis. It provides a more complete picture that illustrates various outcomes of the project. It is one-step ahead in trying to quantify certain parameters that would help (but not substitute!) in making subjective judgment on the project.

Base Case and Assumptions

For the purpose of illustration, the base case for this study is given in Exhibit 2-6. This base case has a couple of deviations from the Excel sheet *Cash Flow Projected Sept. 93* used in Question 1. The initial traffic volume has been assumed to be 20,000 per day and for simplicity, the two improvements are fixed in year 13 and 23 <u>regardless of the actual traffic simulated</u>. Realistically, the timing to initiate these improvements would depend on the actual traffic growth rate, which varies with each simulation. As shown, the base case NPV is negative $175.63 million.

Exhibit 2-6   Dulles Base Case Analysis

Input Parameters for Simulation

The following three key factors have been selected as input parameters for simulation purposes:

(i) The **Initial Traffic Volume** is represented by a normal distribution with a mean of 20,000 per day and a standard deviation of 6,098 per day (statistically this means that a confidence level of 90% is achieved for the initial traffic to stay within the range of 10,000 and 30,000 per day). A plot of this distribution is given in Exhibit 2-7.

Exhibit 2-7   *Traffic Volume Distribution*

(ii) The **Rate of Increase in Toll/Vehicle** is represented by a series of discrete probabilistic values associated with the following rates of increase in toll/vehicle:

Rate of Increase in Toll/Vehicle	Corresponding Discrete Probability
1.0%	0.05
1.5%	0.15
2.0%	0.20
2.5%	0.40
3.0%	0.15
3.5%	0.05

A plot of this distribution is given in Exhibit 2-8.

Exhibit 2-8   *Toll/Vehicle Increase Distribution*

(iii) The **Annual Traffic Growth Rate** are represented by triangular distributions:

- Minimum of 0%, maximum of 15% and mode of 10% for Years 3-14;

- Minimum of 0%, maximum of 6% and mode of 4% for Years 15-22.

For illustration, a plot of this type of distributions for the growth rate in Years 3-14 is given in Exhibit 2-9.

Exhibit 2-9   *Growth Rate Distributions for Years 3-14*

Simulation Results[13]

The simulation results for NPV of the project with respect to each of the three input parameters above are given in Exhibits 2-10 through 2-12, respectively. The results are presented in the form of cumulative distributions of the NPV.

Exhibit 2-10	*NPV Simulation Results*
Exhibit 2-11	*NPV Simulation Results*
Exhibit 2-12	*NPV Simulation Results*

Exhibit 2-13 gives the simulation results for the aggregate influence of all three input parameters. Exhibit 2-14 is the "Tornado" diagram that provides a quick interpretation of the sensitivity of the NPV with respect to each of the input parameters.

Exhibit 2-13	*Simulation Results for all Three Input Parameters*
Exhibit 2-14	*"Tornado" Diagram*

Interpret all these simulation results, relating your argument to the single value of base case NPV.

Concluding Notes

The cash flow analysis presented in this case was "constructed" from newspaper reports and from documents filed publicly in connection with TRCV's application to construct the Dulles Greenway. The analysis is illustrative of many of the procurement issues that have arisen in connection with the project. Ridership data, forecasts, revenues, discount rates, and the like <u>are not actual</u>. In addition, a number of simplifying assumptions have been made in the analysis, which are too numerous to list. The effect of taxes, for example, is ignored. As noted elsewhere in the case, it is difficult to imagine a more successful delivery process for a complex, environmentally sensitive interstate highway type facility. The purpose of this case is to highlight the difficulties associated with complex procurement processes controlling the ways in which handshakes are made between the public and private sectors in the financing, delivery, and operation of public infrastructure facilities. The Dulles Greenway project is illustrative of the need for fundamental, streamlining adjustments in this process, and certainly not indicative of an inability on the part of the EPC sector to deliver high quality, projects on time and within budget.

References

"Dulles Toll Road Runs on Marching Orders." Loudoun Times Mirror December 22, 1993.

Williams, Charles E. "A Road for Today, A Vision for the Future." Construction Business Review March/April 1994.

Dalavai, Leona P. "Impact of the Environment on Construction." The NAWIC Image April, 1994.

Smith, Leef. "Toll Booths To Become Automated." Washington Post July 21, 1994.

Hardcastle, James R. "A 326 Million Private Toll Road to Spur Growth." New York Times July 24, 1994.

Sigmund, Pete. "World's Smartest Highway Under Construction." Construction Equipment Guide November 2, 1994.

Abramson, Rudy. "New Toll Greenway to Ease DC Commute." Los Angeles Times.

Farmer, Tim. "Toll Road Work in High Gear." The Washington Post December 16, 1995.

Pae, Peter. "Drivers Put the Brake on Toll Road's Promise." The Washington Post December 26, 1995, Pages A1, A18, and A19.

"Light Traffic Chills Dulles Debut." ENR January 1/8, 1996.

Resnick, Amy B. "Troubles of Virginia's Dulles Greenway Raise Questions for Private Toll Roads." The Bond Buyer February 16, 1996.

Bailey, Elizabeth. "Driving Up the Learning Curve." <u>Infrastructure Finance</u> July/August, 1996.

"Higher Tolls in the Works for Drivers on Struggling Dulles Greenway." <u>The Washington Post</u> March 29, 1997, Section B, Page 3, Col.1.

Reid, Alice and Spencer S. Hsu. "Private Group Proposes Metro-Dulles Rail Link, Partnership World Fund, Build, Run Connector." <u>The Washington Post</u> May 17, 1997, Page 1.

Notes

[13] Examples of Monte Carlo simulation software packages are At Risk (see www.palisade.com) and Crystal Ball (see www.decisioneering.com).

CHAPTER 3   INTERNATIONAL ARRIVALS BUILDING AT JOHN F. KENNEDY INTERNATIONAL AIRPORT

Infrastructure Development Systems IDS-98-I-201

Research Assistant Melissa Huang prepared this case under the supervision of Professor John B. Miller as the basis for class discussion, and not to illustrate either effective or ineffective handling of infrastructure development related issues. Data presented in the case has been intentionally altered to simplify, focus, and to preserve individual confidentiality.

Introduction

Eleanor Kilcrease, Lead Financial Analyst for the Port Authority of New York and New Jersey, was ready to sleep for a week. Just reaching home at 3:00 AM on May 12, 1997, her mind wandered peacefully through the details of the project she had just finished. For five years, she had worked exclusively on one project, the expansion of the International Arrivals Building in New York's JFK International Airport. Tomorrow, the keys would literally be turned over to the winning Dutch-US management team of Schipol and LCOR. Schipol/LCOR had agreed with the Port Authority to operate the existing terminal during construction of the new $1.2 Billion IAB. As the new project took shape, the Schipol team would demolish the old structure, and then operate, maintain, and finance the new facility over a twenty-five year concession period. Five years of hard work was over.

But . . . . . seeds of doubt remained as to several of the critical financial decisions. As she tried to sleep, her mind inevitably reviewed the project from the very beginning.

History of the John F. Kennedy International Airport

The John F. Kennedy International Airport, located in the southeastern section of Queens County, New York City, on Jamaica Bay, is fifteen miles by highway from midtown Manhattan. Equivalent in size to all of Manhattan Island from 42nd Street South to the Battery, JFK consists of 4930 acres. The airport provides employment for approximately 35,000 people. Exhibit 3-1 shows the general layout of the airport and shows where the IAB is located.

Exhibit 3-1   Map Locating the IAB at JFK

Much of the original master planning of JFK arose from the circumstances and standards of air travel in those first post-war years, not only in technical and functional terms, but also in terms of the public perception. In 1957, going to the airport was an event, one that justified the fountains, expansive plazas and vistas that adorned the original International Arrivals Building (IAB). Air travel was glamorous – airports exciting destinations in themselves. The IAB was the keystone of the eight separate unit terminals that composed the airport. Sophisticated modern buildings expressed the wonder of air travel, where those waiting could watch through a clear glass wall, since made opaque, as passengers proceeded through customs.

The IAB's "elegant efficiency," was challenged over the decades, as the terminal was modified and reshaped in response to industry changes. Large 747 aircraft bringing larger passenger loads heightened security requirements and an ever-increasing annual growth in passenger volume all required alterations to the basic plan. Outside the terminal, landscaped open spaces gave way to parking lots, parking garages, and new roadways. Expansion of the IAB extended its life but altered the original plan beyond recognition.

Earlier Plans for International Arrivals Building

A decade ago, the Port Authority of New York and New Jersey (PA) almost committed to erect a brand-new IAB, one inspired by Grand Central

Railroad Station in Mid-Town. Features of the proposed new IAB included a 115-foot-high, domed, skylight-studded hall. The plan then was that a new IAB would serve as a hub for JFK 2000, distributing passengers to outlying gates via automated "people-movers." The idea, recalls architect Henry Cobb of Pei Cobb Freed & Partners, was to replace the 1957 terminal with a "celebratory space." The proposal went nowhere. The economy soured, some key airlines went bankrupt, and political support evaporated. The PA officially pulled the plug on the idea in 1990.

History of the Port Authority

The PA was established in 1921 by agreement between the states of New York and New Jersey, with the mission of promoting commerce in the bistate port district. The New York-New Jersey metropolitan region consists of the five New York City boroughs of Manhattan, Brooklyn, Queens, Richmond (Staten Island), and the Bronx; the four suburban New York counties of Nassau, Suffolk, Rockland and Westchester; and the eight northern New Jersey counties of Bergen, Essex, Hudson, Middlesex, Morris, Passaic, Somerset, and Union. The PA is a financially self-supporting public agency that receives no tax revenues from any state or local jurisdiction and has no power to tax. It relies almost entirely on revenues – tolls, fees, and rents. In the 75 years of its existence, it has contributed mightily to the construction and operation of the area's infrastructure. Besides JFK, LaGuardia and Newark airports, the facilities it operates include bridges, trans-Hudson tunnels, container ports, industrial parks, bus terminals, and the World Trade Center.

New Plans for International Arrivals Building

In the early 1990's, the PA again determined that the 35-year-old terminal was inadequate to meet the continuing growth in international travel. The IAB was functionally obsolete. Everything in the airline business had changed dramatically -- the IAB had evolved, but not near enough. Space usage, waiting areas, service areas, security areas, and commercial areas simply had not adapted to 35 years of changes in how passengers use and move through airport terminals. While the structural systems were sound and clean, the numerous deficiencies were readily apparent. An assessment of the IAB terminal in relation to international terminal (IATA) criteria and standards, listed the following faults (from the Request for Proposal):

- Disorientation due to building layout and environment
- Long passenger processing times
- Long walking distances
- Inefficient handling of passengers, well wishers, and meeters/greeters
- Overcrowding at peak travel times
- Circuitous circulation
- Multiple security points
- Poorly located retail
- Limited gate flexibility
- Demand greater than capacity
- The 1957 Air Handling Units (AHUs) are in poor condition and the 1970 AHUs are also in deteriorated condition.
- Majority of fire and smoke detection systems in need of upgrade
- No centralized building services controls or monitoring systems
- Communication/information systems out of date; include many manual procedures.

There were more than 40 airlines operating at the IAB, with 50% of the fleet mix composed of Boeing 747 aircraft. The demand pattern is highly peaked, with most flights arriving and departing between 2:00 and 10:00 p.m. Shortcomings on the airside include:

- Not all gates accommodate large aircraft.

- Non-optimal utilization of ramp space and congestion on service roads

- An obsolete and inadequate communications/gate management system.

Working with the airlines, the PA undertook a joint effort to redevelop the IAB. Among the 14 signatory Tenants there were divergent proposals to address the IAB problems, ranging from a major refurbishing to total reconstruction. As each option was explored, it became clear that the price tag would range from $600 million to $1.1 billion. As the Tenants discussed their business options, four Tenants formed a consortium, leased the vacant Eastern Air Lines Terminal, and announced a $430 million investment to build a new 11-gate terminal to be called "Terminal One". While the remaining IAB Tenants regrouped, the Authority took the lead and began preliminary feasibility studies for the IAB.

Terminal One

Air France, Japan Airlines, Korean Air, and Lufthansa are equal partners in the $435 million Terminal One project. It will be their new home when it opens in 1998, the first new terminal at JFK in more than twenty years. The Terminal One Group Association (TOGA), a limited partnership comprised of the four airlines named above, signed a lease for the 36-acre site with the PA in 1994, with financing arranged through the New York City Development Agency. TOGA formed Terminal One Management Inc. (TOMI) to manage financing, construction and operation of the terminal. Each of the airlines is represented in TOMI's executive suite.

Feasibility Analysis of the IAB

If isolated as a stand-alone terminal, the existing IAB would be the fourth largest international airport in the U.S., serving 45 airlines and over 6 million passengers per year from 14 gates. Currently, it is the only terminal at JFK still operated by the PA. As the owner-operator of the IAB, the challenge to the Authority was to balance two principal needs – the physical requirements of an aging and obsolete building and the financial requirement to preserve the cash flow from the facility at a reasonable pass-through cost to the airlines that use it. An A/E team and a forecasting and financial feasibility modeling team were quickly selected to assist in making the best business decisions.

The cost increments from the feasibility analysis were staggering. For $200 million, the building systems could be renovated; for $600 million, modest improvements in the ticketing areas and gate availability would also be achieved, but the traveling public would see few internal amenities. At $1.0 billion, complete demolition and construction of a new terminal plus apron and frontage roadway replacement could be accomplished.

It was at this point in time that Eleanor was assigned to the IAB Project. Her first task was to do an analysis of the project and to give recommendations on how to proceed. The financial risks associated with undertaking a project of this magnitude raised key issues:

- What would it take – and what would it mean – to get long-term commitments from prospective airline tenants? How firm would these commitments be, given the volatility of the aviation industry? How firm did they have to be?

- What passenger volume would be required – and what level of confidence does the forecast hold – to permit the project to proceed based only on per passenger charges and a per-use tariff to the users, i.e. with no long-term airline commitments?

  - Other variables were highlighted in the risk analysis:

Careful competitive analysis led the PA to focus on the assumption that there may be no signatory tenants. Supply and demand factors indicated the IAB was essential to the market and a core level of usage could be guaranteed. The metropolitan New York region generates high demand for North Atlantic traffic. Also, there was little risk that operators of the other eight terminals at JFK could expand their international capacity to absorb the IAB's 45% market share. Worst case scenarios left the IAB with half of its current market share of the international activity, assuming major investments by airline alliances at the other terminals took place.

Revenue Factors:	NY Region International Market Share of North Atlantic Flights
	JFK International Market Share of NY Region
	IAB International Market Share of JFK
	Passenger Growth Rate
	Revenue per Passenger:
	Enplanement Rate
	Concession
Construction Factors:	Construction Cost
	Construction Schedule
Operational Factors:	Restrictive Covenants
	Operations During Construction
	Industry Changes
Business Factors:	PA Capital Capacity
	Private Industry Capability
	Environmental
	NYC Long Term Lease

Table 3-1   Risk Factor Analysis

Eleanor's feasibility work concluded that JFK's air traffic market demand was sound and that construction of a new terminal was financially feasible and the most desirable option. Net present value (NPV) analysis showed that while complete new construction produced lower cash flow initially, it was, in the long-term, the best alternative with the best return on investment. JFK had problems handling the growth in air traffic. In a few years, the IAB would burst its seams and break down completely. Mere renovations would strengthen the seams but would not solve the problem.

The next obstacle was earmarking $1 billion to construct the terminal. With a desire to accelerate the project and a realization that competing interests might complicate the financing of the project, the Authority simultaneously initiated design and the investigation of alternative financing plans via privatization options.

Preliminary Design - 1993 to 1994

Recognizing that the terminal facility no longer functioned as it was intended, it was decided that a significant capital investment would be made to restore the IAB and the redevelopment objectives were set. (See Exhibit 3-2 on the companion CD) In June 1993, after an extensive competitive process, the A/E team of TAMS Consultants Inc., Skidmore, Owings & Merrill, and Ove Arup & Partners was given 12 months to rapidly advance the preliminary design for the terminal (see Exhibit 3-3). Working with the PA's engineers, the charge was to bring the design to the 25% completion level, produce a Basis of Design Brochure summarizing the project's functional and design criteria and associated drawings for all aspects of the project. Supplemental documents include a geotechnical report, life safety code analysis, construction staging plan (see Exhibit 3-4), a proposed retail plan, contract procurement strategy, an overall cost estimate including construction cost estimates, and an implementation schedule. Separately, an operations and maintenance plan and budget for the new terminal were prepared.

Exhibit 3-2	*IAB Redevelopment Objectives*
Exhibit 3-3	*Preliminary Design*
Exhibit 3-4	*Staging Plan*

A quick commitment to a design was driven by:

- The unacceptability of further time delay

- No guarantee of private sector interest in the project

- A better position for partner selection with a more advanced base plan.

- In the absence of any precedent for an airport project of comparable scope and magnitude, there was concern about the character and quality of the responses that would be answered by an early preliminary design.

The 25% design by the A/E team provided for a 16-gate terminal including facilities for opening two additional gates and provisions for adding 21 more. The design also provides for direct transfer of rail passengers inside the terminal, should transit ever come to New York City's largest airport. This design was based on the projected needs in 2005 with a peak hour capacity of 3200 international deplaning passengers and 2600 international enplaning passengers. The terminal can be expanded to accommodate future IAB traffic with a peak hour capacity of 6000 deplaning passengers per hour.

The Pre-Qualification and Bidding Process - 1995 to 1997

By mid-1995, sufficient project details were available to support a Request for Qualifications (RFQ). The RFQ requested information beyond just company experience and qualifications. By providing a comprehensive briefing book with the RFQ, respondents were asked to identify their financing structure and business relationship with the Authority. They were also asked to estimate enplanement rates, equity participation, returns to the Authority and issues that should be addressed to enable the respondent to make a firm business proposal.

Seven teams were identified as having the qualifications to proceed with the contract process. All seven teams had conflicting sentiments about the project. There was strong interest combined with an equally strong skepticism that the deal would materialize due to the earlier cancelled construction plan by the PA. To deal with this skepticism, in an unprecedented fashion, the PA gave the developers access to all relevant design documents prior to the release of the RFP. Preliminary business terms for the deal were released. Developers were encouraged to submit alternate designs. Two full days were set aside whereby the developers could query the Authority on design, financial, business and operational matters. This open-book approach enabled the PA to demand an equally comprehensive and detailed proposal.

Of the seven prequalified teams, three decided to drop out. These include BAA USA Inc., JFK-IAB Partners (Turner Construction, JP Morgan), and Raytheon Infrastructure Services. Two of the teams cited difficulty in filling out their team and the third questioned the project's feasibility. The four that decided to continue with the process included:

- **IAB Gateway Developers**: Airport Group International, United Infrastructure, Goldman Sachs, Merrill Lynch

- **Idlewild Associates**: Johnson Controls, Lehrer McGovern Bovis, Hines Development, Paine Webber/Bear Stearns

- **Schipol USA/LCOR Incorporated**: Schipol, LCOR, Morse Diesel, Fluor Daniel, Lehman Brothers, Citicorp Securities

- **Ogden Corporation**: Ogden, Tishman Construction, Ralph M. Parsons, Smith Barney

The Authority formed an evaluation team augmented by consultants familiar with the project. O'Brien Kreitzberg, the Program Manager for the overall JFK Redevelopment effort, provided the technical support and expertise. Financing experts from Fullerton & Friar and airport operating experts from Landrum and Brown and NAPA Airport Development Consultants were brought on board. Cambridge Partners was selected as the PA's financial advisor. These different components were broken up into three independent panels – Business and Finance, Development, and Operations and Management. These panels will conduct the initial review of the proposals. The Development panel probed the cost estimate, design concept and construction schedules. Project phasing was of particular importance, since the plans had to accommodate the 6 million annual passengers and 40 airlines using the terminal during the construction. The Operations and Management panel reviewed the depth of the proposer's terminal management knowledge, staffing and cost estimates for running the terminal. The Business and Finance panel closely scrutinized funding sources, flow of funds, equity, debt service coverage and overall cohesiveness of the financing packages, plus airline and retail leasing plans.

With the distribution of the Request for Proposals (RFP), the PA listed their objectives in the entire project (see Exhibit 3-5 on the companion CD). In order to best fulfill these objectives, along with a base proposal using the PA's approach from the preliminary design, proposers were allowed to submit an alternate proposal and design.

Exhibit 3-5   *RFP Criteria*

The PA's objectives include the following:

- To develop a new facility which meets the needs of the airport, the traveling public, and the airlines serving the airport;

- To keep the charges to the airlines using the terminal at reasonable levels which are comparable to the cost of other international facilities at the airport;

- To provide a reliable "baseline" revenue stream for the PA from the facility, while providing the PA an opportunity to share in the "upside" potential of revenues generated in the terminal;

- To minimize or, if possible, eliminate the PA's financial risk related to the development and future operation of the facility, and;

- To minimize or, if possible, eliminate the need for the PA to incur any capital costs for the project or issue debt obligations backed by the credit of the PA.

Four comprehensive proposals were received on March 4, 1996. According to the RFP the proposals were evaluated based on the following criteria:

- **Financial Return to the Port Authority**: Proposals will be evaluated on the valuation of guaranteed and variable payments made to the Authority and the conditions or assumptions which support the payments. The long term return will be examined for value for the initial ten year period and through the 25 year lease term. Interim payments made during construction and prior to DBO will also be examined. A net present value analysis will be the Authority's primary, but not exclusive, basis for evaluating financial return. In addition, an assessment of risk related to the ability of the Authority to realize its return will be conducted;

- **Financial Plan**: Proposals will be evaluated on their level of capital commitment, the demonstrated feasibility of the plan, the financial resources of the Proposer, the sources of capital and the conditions assigned to the proposed plan. The Authority will analyze all estimates of construction and operating costs, enplanement rates, concession revenues and rents, among other factors, in evaluating the financial plan;

- **Development, Management and Operations Plan**: Proposals will be evaluated on the quality of design, the level of service provided to tenants and passengers, construction staging and schedule, and initial and life-cycle capital costs. The management and operations plans will be evaluated for quality of the O&M estimates, retail plan, ramp operations plan, leasing plan and transition planning for operations during construction and for the utilization of current staff in the IAB;

- **Team Experience**: Proposals will be evaluated on the team's experience relating to capital formation, international air terminal design, terminal construction, and terminal and aeronautical management and operations;

- **Overall Quality and Cohesiveness of Proposal**: Proposals will be evaluated on the responsiveness, feasibility and overall content of their proposal. The level of integration and coordination of individual components of the proposal will also be considered.

Six weeks later, the PA had a signed Memorandum of Understanding (MOU) covering all basic business terms with the selected team of Schipol

USA/LCOR. The selected design was a variation of the Authority's original plan, with slimmer concourses and an expanded and upgraded retail court. The largest public/private airport deal in the United States was underway.

New York Land Lease Problem

As Eleanor discovered in her preliminary analysis, a key future problem facing the private operators is the expiration of the land lease with New York City (NYC) on December 31, 2015. Since 1947, the PA has been leasing all 4930 acres of airport land from New York City. Because of its location just 15 miles from midtown Manhattan, the land is presumed by many to have a high market value. The lease requires PA to pay rent equal to the net revenue generated by operations at JFK, with a minimum guaranteed annual rent of $3.5 million.   There has been an ongoing argument between NYC and the PA on whether or not the PA has been shortchanging it on rent, at least since 1991. Payments have dropped from a peak of $80 million in the late 1980's to $6.2 million in 1994 and $14 million in 1995. New York City claims that a cumulative shortfall of $400 to $800 million is owed from the PA in "rent" payments since 1991.

The impact of this conflict on the IAB project will be unknown until the termination of the current lease with NYC in 2015. The city then has the option of renting out the land at a much higher price or could possibly even decide not to continue to lease the land. The latter possibility is considered to be very unlikely since the airport generates major economic benefits by providing over 173,000 jobs through on- and off-airport aviation and indirectly related businesses.   JFK contributes $15.8 billion annually in economic activity to the NY/NJ region, of which $4.8 billion is in wages and salaries. To deal with this issue, the PA asked proposers in the RFP to identify their assumptions as to the city's actions at the end of the lease.

Consortium Members - JFK International Air Terminal LLC

JFK International Air Terminal LLC (JFKIAT) combines the resources of LCOR Incorporated, an accomplished national real estate firm; Schipol USA, the American affiliate of the firm that operates the widely acclaimed Schipol airport in Amsterdam; and Lehman Brothers JFK, an affiliate of Lehman Brothers Inc. In addition, JFKIAT includes Fluor Daniel, Inc., as construction program manager; Morse Diesel International as construction manager; and TAMS Consultants, Skidmore, Owings & Merrill and Ove Arup & Partners as architects of the new terminal, and Communications Arts as retail designer. Lehman Brothers, Inc., and Citicorp Securities, Inc., serve as the project's financial advisors.

Charles A. Gargano, Vice-Chairman of the PA, said, "This project is another important step in the privatization of PA facilities. We reached out to the private sector to draw on its expertise in delivering top-quality services to our customers. The new terminal will accommodate millions of travelers a year and strengthen the region's commanding position in tourism and international business. The winning team, JFKIAT, includes firms recognized around the world for their expertise."

The Amsterdam Airport Model

Schipol USA is a subsidiary of the company that owns and operates Amsterdam's exceptionally functional airport. Schipol Airport has led Business Traveller magazine's poll as best European airport since the early 1980's. Only Singapore/Changi ranks higher on world listings. U.S. airport terminals are not ranked highly. The Amsterdam airport is less revered for its architectural distinction than for its hyperefficiency at getting passengers on their way *and* at lightening their wallets. The Dutch have exploited the potential of airports as shopping plazas for passengers with time and money on their hands, as testified by Schipol's glittering shops. Unable to resist the temptations of the airport's consumer amenities – which include a full mall with designer boutiques and a casino – travelers spend an average of $35 per trip, compared with $17.50 a head in JFK. The airport also sets maximums of 15 minutes check-in time and just six minutes for immigration. The Amsterdam airport is painless and profitable, a combination for which the Port Authority and the new proprietors of Terminal 4 hunger. Accordingly, the new IAB terminal will fulfill a pragmatic vision of a commercial hub that comforts travelers by immersing them in a familiar environment: a shopping mall, and a universe of small things in the Center Retail Court. This mall, which naturally has a New York City theme, is filled with wacky, freeform fixtures (some of which double as heating and air-conditioning ducts).

Financing Strategy

The financing package proposed by Schipol was a massive undertaking which is anchored by a loan of $932 million for construction costs. The IAB project was to be a pure project financing, with no recourse: to the airlines that will use it, to the three companies that signed the 25-year lease to build and operate it, or to the PA. Prepayment of the debt is secured solely by the revenue stream associated with terminal operations. There are three keys to the success of the off-balance sheet financing:

1.　　PA was willing to share control of a major profit center with a private developer/operator.

2.　　PA agreed to let The Bank of New York, the bondholders' trustee, control the selection of a new operator if a serious problem arises with Schipol. Because of this provision, Ernest Perez, airport analyst with Standard & Poor's, says, "The Port Authority hasn't really lost control of this facility. If something goes wrong with the transaction, the Port Authority will step in."

3.　　PA agreed to make $80 million in IAB access improvements and subordinated most of its share of the net profits to bondholders.

Of the entire required costs of $1.2 billion, the majority ($932 million) was provided through special project bonds backed solely by airport revenues. Even with financing backed solely by terminal revenues, the project received an investment-grade rating from three agencies, and qualified for bond insurance (see the Fitch report, Exhibit 3-6 on the companion CD). The essential nature of the project translated into demonstrable demand and market-share potential. With the bonds' ratings of a BBB+ from Standard & Poor's, Baa2 from Moody's, and an A rating from Fitch, Monoline insurer MBIA Inc. guaranteed the debt service with an AAA-rating and the IAB project was truly underway. On April 25, 1997, the special project bonds were sold to institutional investors. The largest airport bond issue ever and the first major airport privatization in the U.S. sold out in less than 90 minutes.

Exhibit 3-6   The Fitch Report

This means that the financing contains only a token amount of parent company equity - $15 million: 40% from Schipol; 40% from LCOR; and 20% from Lehman. Of the total, $10 million is to be placed in a lease contingency reserve during the construction period. The remaining $5 million will pay part of the construction cost. In handling the revenue stream, the PA accepted a 60/40 split (60% to the PA, 40% to the private operators) of the net revenues after operation and maintenance expenses and debt services are paid. (See Exhibit 3-7 on the companion CD) However, the PA agreed to cap its share of the revenue stream at $60 million a year. The PA also allowed the operators to take their management fee up front as part of the operation and maintenance costs. In return, the operators pay a guaranteed base rent of $12 million a year to the PA during the construction period. After that time, the $12 million is set as the minimum terminal rent to the PA.

Exhibit 3-7   Historical and Predicted Passenger Volume

Closing the Deal

In May 1997, with over $30 million already committed by the developer, the lease was signed, JFKIAT took the keys to the existing terminal, construction trailers moved in, and site work began. The new IAB terminal is expected to open in 2001, with the demolition of the old terminal completed in stages as the new terminal takes its place. The PA's goals from the onset of the search for a private sector partner had been achieved. They were:

- Construction of a new terminal

- Competitive rates for airlines

- Reliable revenue base and upside sharing potential

- Appropriate risk sharing

- Minimal contribution of capital and no PA-backed debt.

Despite these advantages, Eleanor had argued against Schipol; she believed that the financing package Schipol proposed, while ingenious, did not provide enough of the financial upside from the project to the PA. She felt the PA could have gotten a bigger share of the profits. Alternatively, she thought that the PA could have added several small projects to the IAB development, including the approach roads. Even with these additional features included, Eleanor thought that the rate of return on the project would have been sufficiently attractive to all the competitors.

As a relatively new player in the mix, her objections were largely ignored, and Eleanor had let the matter rest and the contract was signed.

Questions

Suddenly, Eleanor was wide awake -- her doubts screaming at her. Should she have made a bigger fuss? Help Eleanor settle a number of questions in her mind so she can get a good night's sleep tomorrow.

1. Where does this project fit in the Quadrant framework? What factors peculiar to this project made this procurement strategy possible?

2. Review the Cash Flow Analysis presented in Exhibit 3-8 and construct your estimate of Schipol's expected total NPV. Does the project appear profitable from Schipol's perspective? Do you think the Port Authority appropriately leveraged the value in the project for public gain?

Exhibit 3-8   Cash Flow Base Case

Make the following assumptions:

a. The first installment of bonds are equally divided among maturing years so that $27,461,538.46 worth of bonds mature each year of 2003-2015 at a yield rate of 5.485%. Model the maturity and yield of remaining installments according to the "Financing Package" sheet;

b. The net effect of taxes is 25% annually (this accounts for tax expenses net of any tax shields);

c. Neglect inflation (i.e., cash flows are treated in real terms);

d. Initial O&M expenses are $100 million and grow in real terms at 4% per year. Ground rental charges are as shown through year 2005. Assume that subsequent ground rental charges grow at 4% per year for the remaining periods of analysis.

3. Conduct a sensitivity analysis by calculating the adjusted total NPV given the following scenarios. Compare each of these scenarios with the base case from Question 1. Show your work on separate worksheets.

What if construction costs overrun by 10%? decrease by 10%?

What if revenues per passenger increase by 10%? decrease by 10%?

What if the passenger volume increases by 10%? decrease by 10%?

What if O&M costs increase to 10%, instead of 4%?

4. What if New York City decides that it wants a fixed percentage of the profit as the terms for the new lease to be signed at the end of 2015? What if this percentage is fixed at 20% of net income? How would this affect Schipol earnings?
5. The Schipol contract, signed in May 1997, provides a construction period of only three years. The operating side of the contract with

Schipol (the 25 year lease) begins upon the completion of construction (i.e. a total contract time of 28 years from 1997-2025). Would the NPV be affected if construction time (but not the price) overran by 1 year? Under ran 1 year?

What changes do you recommend in the lease terms in order to provide additional incentive to Schipol to finish early, within budget?

6. Based upon your projections from Question 2, at what expected revenue percentage does Schipol just cover its investment at the end of the contract?

7. If you were in Eleanor's shoes, would you have pressed for a higher share of profits for the Port Authority? What percentage seems reasonable to you, in all of the circumstances?

Can Eleanor get a peaceful night of sleep tomorrow?

CHAPTER 4   THE SR 91 EXPRESS LANES

Infrastructure Development Systems IDS-97-T-012

Research Assistant Om P. Agarwal prepared this case under the supervision of Professor John B. Miller as the basis for class discussion, and not to illustrate either effective or ineffective handling of infrastructure development related issues. The facts in this case have been altered for the purpose of considering the procurement strategy issues in an educational setting. The assistance of Roy W. Nagy of Caltrans' Office of Public /Private Patrnerships in the preparation of this case is gratefully acknowledged

The Problem

In the mid-eighties, the California Department of Transportation (CAL DOT) also known as Caltrans, faced a problem that has become common throughout state DOT's across America: how to pay for a growing list of badly needed highway capital projects, while capital resources dwindled and existing maintenance requirements grew.

Beginning with the enactment of the Interstate Highway System legislation in 1956, highway construction and maintenance has been financed through a combination of dedicated gasoline taxes, motor vehicle registration fees, and direct federal aid. While this combination had proved effective for over a quarter century, the anti-tax movement, particularly in California, had made raising additional taxes very difficult. The need for improving, rehabilitating, maintaining, and, indeed, expanding California's road system continued. (See Exhibit 4-1 on the companion CD)

Exhibit 4-1   Financial Analysis

In the late eighties, a variety of possible methods for sustainable financing of road projects in California were being debated, including additional bond authorizations, the creation and levy of so-called "impact fees" on real estate developers, and sales taxes for transport improvements. In 1988, Bob Poole, of the Reason Foundation[14], proposed that the private sector might build private toll roads to fill the gap between the level of services demanded by the public and the public's apparent unwillingness to permit the government to do so directly through additional tax or use charges. For additional background, see Calleo.[15] Following a conference in August, 1988, at which a number of developers made presentations describing the potential benefits that alternate delivery methods might provide to meet the State's unmet transportation needs, a number of California groups, with the cooperation of Caltrans, its then Director Robert Best, and its then Assistant Director Carl Williams successfully encouraged the California legislature, in Assembly Bill 680[16] to authorize Caltrans to solicit proposals and enter agreements with private entities for the construction, lease, and operation of up to four public transportation demonstration projects. (See Exhibit 4-2 for a summary of AB 680)

Exhibit 4-2   Summary of AB 680 Legislation

Key Features of AB 680

The Bill sought to strike a balance among the concerns expressed to Caltrans when the concept of private toll roads was being debated, including the wages that private developers would pay for labor. Proponents wanted the Bill to have broad public support and acceptance, yet it had to attract the interest of private developers and investors as well. Some of the elements included in the Bill to strike this balance were these:

1. At least one project of the four had to be selected from either the Northern or Southern end of the State. This was believed necessary to secure the support of a larger number of legislators in both the San Francisco and Los Angeles dominated regions of California.

2. Each project would have to be a supplement to existing facilities. In other words, none of the four selected projects could comprise an exclusive transportation service, for which there wasn't a non-tolled public alternative.

3. Each of the four demonstration projects would be owned by the State at all times. This feature, it was thought, would have tax advantages for

potential developers, and would reduce the liability risk of operating a public transportation facility to private operating companies.

4. Each selected project would be leased to the Developer for a term of 35 years, which it was hoped, would enable each private operator to recover its capital and operating investment <u>and</u> earn a reasonable return.

5. Each selected project would have to be financially self-sufficient, with <u>no</u> State or Federal funds required or permitted to be invested.

6. Each project developer would fully reimburse the State for any services that it would be required to provide, such as highway patrol or maintenance services.

7. Each private developer would have authority to impose and collect tolls sufficient to recover its costs and secure a reasonable return. The Bill was silent on what constitutes such a return.

8. Each project agreement would require that "excess toll revenues" would be applied towards early repayment of private sector debt for the facility, or paid into a State Highway Account.

9. The State would retain the right to charge tolls after expiry of the lease to recover the State's O&M costs, once it took over the facility.

10. Plans and specifications for each project would be required to meet or exceed State design standards, and to comply with other State laws and regulations, such as environmental and local land use regulations.

11. The projects would be treated as Public Works and would, therefore, come under the purview of the labor acts.[17]

The History of Private Toll Roads in the United States

Private toll roads are not a new concept in the United States. Throughout much of the first 150 years of the American republic, numerous private toll roads had been developed, often with the indirect support of the federal government. Such support took the form of alternating land grants, contracts to carry the mail over specific post roads, and federal surveys conducted by the Corps of Engineers.[18] In fact, many public roads were built as supplements to private toll roads. The emergence of the railroad engine as a high quality provider of transportation shifted federal incentives away from private toll road construction toward railroad construction in the mid 19th century. Railroad construction was similarly delivered through private sector finance, with alternating land grants, contracts to carry the mail, and federal surveys as incentives.[19]

The 20th century saw the need for a good trunk road system, particularly after the advent of the internal combustion engine. The Federal Highways Act of 1956 authorized a comprehensive Inter-State highway system, to be largely funded through 90% federal subsidizing grants to reimburse states for design and construction costs. A federal gas tax and a federal excise tax on motor vehicles and parts was levied to provide a dedicated source of revenue to fund the new federal capital program for interstate highways. With the advent of roads paid for by the federal government, state funding for state limited access highways was quickly curtailed.

CALTRANS' Pre-qualification Process

The increasing need for highway capacity, coupled with a constrained capital resource base, produced interest in private toll roads. Once AB 680's demonstration program became law, Caltrans lost no time in taking advantage of this new opportunity. A 'Privatization Advisory Steering Committee' was set up under Carl Williams and after a first stage of screening, ten potential developers were pre-qualified by Caltrans to submit 'Conceptual Proposals'. The developers were pre-qualified based on criteria in the Request for Qualifications. See Table 4-1.

The ten developers pre-qualified by Caltrans were:

1. Bechtel (by itself)
2. A consortium led by Bechtel
3. A consortium led by Ebasco
4. A consortium led by Perini Corporation
5. A consortium led by Wright from New Zealand
6. A consortium led by T.Y Lin from San Francisco

7. A consortium led by Parsons Corporation
8. A consortium led by CRSS Commercial Group
9. A consortium led by Parsons Brinckerhoff
10. A consortium led by Ross Perot Jr. of Dallas and Greiner.

Pre-Qualification Criteria	%Weight
Experience of the principal organization and consortium members	30 %
Record of financial strength to commit to a major transportation facility	30 %
Ability to work cooperatively with a broad range of governmental agencies and the public	20 %
Individual qualifications of key project team personnel	10 %
Organizational and management approach for project company or consortium	5 %
Familiarity and experience with automated traffic operations, Automatic Vehicle Identification (AVI) and Electronic Toll Collection systems	5 %

Table 4-1   Pre-Qualification Criteria
Source: Caltrans RFQ, November, 1989

Nine of the ten developers were teams of firms with different backgrounds and skills, but which together, provided the transportation, technology, design, construction, financing, and government experience Caltrans sought in its Request for Qualifications. The collection of pre-qualified developers represented an entirely different delivery and finance approach for designing, financing, constructing, and operating public infrastructure facilities, and a marked departure from statutory segmentation of planning, design, construction, finance, and operators that began with the IHS program 40 years earlier.

The Call for Competitive Conceptual Proposals

In the second stage of Caltrans' procurements under AB 680, 'Conceptual Proposals' were requested from each consortium. The proposers, not Caltrans, made the choice of transportation projects. The idea was to minimize expenditures by the private and public sector on the details of the particular projects prior to the time that actual proposals were selected for implementation, yet to provide sufficient information to permit a fair comparison and evaluation of competing proposals. Caltrans issued a set of guidelines for submission of conceptual proposals in March 1990.

Among the terms and conditions established in the guidelines was that each proposed project must be implemented at no cost to the State. These guidelines established weighted selection criteria in nine categories.

Evaluation Criteria	Points
Transportation service provided as a result of the proposal.	20
Degree to which proposal encourages economic prosperity and makes overall good business sense.	10
Degree of local support for proposal.	15
Relative ease of proposal implementation.	15
Relative experience and expertise of the proposal sponsors and their support team on similar projects.	15
Degree to which the proposal supports the State's environment quality and energy conservation goals.	10
Degree to which non-toll revenues support proposal costs.	5
Degree of technical innovation associated with the proposal.	10
Degree of proposal's support for achieving the civil rights objectives of the State regarding the utilization of Minority and Women Business Enterprise.	10

Table 4-2   Evaluation Criteria
Source: Caltrans RFQ, March, 1990

The Proposals

Eight proposals were received in response to Caltrans Call, including one submitted by the California Private Transportation Company (CPTC), a company formed by the CRSS group, for the addition of toll lanes in the median strip of SR 91. Exhibits 4-3 and 4-4 are graphical representations of the project and are included on the companion CD.

Exhibit 4-3   *Toll Road Entrances*
Exhibit 4-4   *Toll Zone*

Brief History of SR 91

SR 91, which opened in 1968, is an 8 lane east-west inter-state highway starting from SR 110 in Los Angeles County and running across Orange County into Riverside County. It is a principal link between the coastal and inland regions, traversing through the Santa Ana Canyon between the Cleveland National Forest and the Chino Hills. This freeway has seen a rapid rise in traffic volumes, as it is a major commuter link between the residential and employment centers in Orange, Riverside and San Bernadino counties. Traffic has increased from 91,000 vehicles per day in 1980 to 188,000 vehicles in 1990. Traffic volumes are estimated at 255,000 vehicles per day in 1995. The proposed developer of the SR 91 toll lanes, CPTC, estimated traffic volumes of 330,000 to 400,000 vehicles per day by 2010.

There was a perceived urgent need for enhancing the capacity of SR 91 between Orange County and Riverside County. In addition to lack of state funds, obtaining an appropriate additional right of way seemed insurmountable, given the nature of the terrain and the high land costs in the already developed areas adjacent to existing SR 91. Earlier studies had estimated that the addition of HOV lanes would be a low cost option and more realistic. There would be a need to relax existing standards of lane, shoulder and median widths in order to accommodate additional capacity. This would be easier if the additional capacity was in the form of HOV lanes.

The two counties and Caltrans had signed an agreement in 1988 to jointly fund the planning and design of these HOV lanes. Orange County had not, however, been able to secure support for a sales tax hike needed to fund such a study. They had made a request for inclusion of this project in the next State Transportation Investment Plan (STIP) so that State funding could be secured. The California Transportation Commission had approved the project, but planning was not slated to start till 1996.

CPTC's proposal to develop the median area of the expressway as a private toll road fit into this long term planning strategy.[20]

Key Features of the Proposed SR 91 Toll Expressway

CPTC's proposal was for a 10 mile long 4 - lane facility in the median of SR 91. While not an exclusive HOV facility, it would provide free access to HOVs as a measure of implementing State objectives. CPTC's proposal stated that the project would cost $ 126 million and could be completed in less than 3 years. Included with the proposal was Orange County's offer of a $ 20 Million loan to help finance the project. For background information on the SR91 project, see References and the end of the chapter and Exhibits 4-1 through 4-7, on the companion CD.

Exhibit 4-5	*Application Agreement*
Exhibit 4-6	*Toll Schedule*
Exhibit 4-7	*Backgrounder: Technology*

Consortium Members

The sponsor, California Private Transportation Company (CPTC) is a limited partnership between Peter Kiewit Sons' Inc., Cofiroute Corporation and Granite Construction Inc.

Peter Kiewit Sons' was established in 1884. Its business is construction, mining, telecommunication, energy and infrastructure industries. It is the major investor in CPTC and provided project management, construction management and financial services for the project.

Cofiroute Corporation is the American subsidiary of a French toll road operator, Compagnie Financiere et Industrielle des Autoroutes (COFIROUTE), focusing upon private tollway development and operation.

It provides assistance and advice in the areas of operations, electronic toll collection and traffic management for the SR 91 project.

Granite Construction Inc., established in 1922, is a large transportation contractor in the US. Granite is also a partner in a joint venture led by the Kiewit Construction Group for the Design-Build delivery of the Orange County Transportation Corridor Agency's San Joaquin Hills toll road.

A then subsidiary of Kiewit, MFS Communications Company Inc. (MFS) was contracted to provide AVI tolling equipment for the project. (See Exhibit 4-8 for partner information.)

Exhibit 4-8   *Partner Fact Sheets*

The Proposed Development Franchise Agreement

CPTC proposed a Development Franchise Agreement with the following general characteristics. **For purposes of this case, the actual agreement signed with CPTC for the SR 91 project is treated as a draft.** Key provisions of the Draft Agreement are summarized below:[21]

Facility
Approximately ten mile segment of up to 4 buffer separated, express lanes of roadway on the median between Caltrans' east bound and west bound lanes of State Route 91, between the initial eastern terminus at the boundary line separating Orange county and Riverside county and an initial western terminus west of the interchange of State route 91 with State route 55 in Orange county. (See Exhibit 4-9 for Caltran project rankings.)

Exhibit 4-9   *Caltran's Project Rankings*

Rights granted
For the design, development, acquisition, construction, installation and operation of the project.
Exclusivity
No similar franchise to be granted to any other party within a pre-defined 'Absolute Protection Zone'.

Franchise fee
Base fee of $ 10 per month.
Variable Franchise fee
50% of available cash flow can be retained by CPTC as incentive
    return and remaining 50% to be paid to Caltrans as variable
    franchise fee whenever Base NPV>0 and Total NPV<0.
Excess Franchise fee
All toll revenues in excess of the permissible Return on Investment
    to be paid into a State Highway Account. These are the
    available cash flows when Total NPV is = or > 0.
Base Return on Investment
Base Return on Investment allowed is 17%.
Incentive Return on Investment
Linked to the increase in Annual Peak Hour Vehicle Occupant
    Volume (APHVOV) and is 0.20% for every 1% increase in the
    APHVOV, subject to a ceiling of 6%.
Reserve Funds
Limits imposed on amounts to be kept on various reserve funds.
Reports
Annual reports to show the revenue flows and to be used for
    determining the variable/ excess franchise fees and the Base
    return and Incentive return.
Control on toll rates
CPTC free to decide the toll rates. Only returns are subject to limits.
Design and Construction Standards
As per State standards. Caltrans to review design and plans before
    commencement of construction work
Environment Clearance
Responsibility of CPTC to obtain this. Final approval of facility
    contingent upon completion of environmental review under the
    California environmental statutes.

Safety of operating practices
Traffic operating plan to be got approved from Caltrans.
Land Acquisition
CPTC to negotiate and secure private parcels. Costs to be
    incorporated as part of Capital costs. Facility of condemnation
    available through Caltrans.

Toll enforcement and penalties

Services of the California Highway Patrol to be retained at a
  mutually agreed price for toll enforcement. Provisions of the
  California Streets and Highways Code to cover this facility for
  penalizing violation.

Third party claims

Elaborate allocation of responsibility between CPTC and Caltrans.

Financing Package

CPTC's proposal was to fund the $126 Million project as follows:

Source	$$ Millions
Funded Equity	$19
Fourteen and one half year term loans from Citicorp, Banque Nationale de Paris & Societe Generale	$65
Institutional Tranche purchased by Peter Kiewit Sons	$35
Three year post completion subordinated loan at 9% from Orange County Transportation Agency	$7

Table 4-3   CPTC's Funding Proposal

Matt Moore's Tasks

Matt Moore is one of four members of the headquarters Caltrans
evaluation team that is reviewing the proposals submitted pursuant to AB
680.

Review teams in the districts where each of the proposed projects were
located had first evaluated each project. These reviews were conducted in
locked rooms and under very specific instructions that neither the proposals
nor the ongoing evaluations were to be discussed with anyone not on the
district review teams.

The district evaluations were presented (orally and in writing) to the headquarters review team in the first Headquarters team meeting several weeks ago. At that meeting, strict instructions were given to everyone involved: (1) don't discuss the process or the issues with anyone not on the headquarters review team, (2) don't take any documents out of the headquarters evaluation room, and (3) don't discuss the issues with anyone unless it is in the evaluation room.

After the initial presentations by the district review teams, Matt and the rest of the headquarters team reviewed the individual proposals in detail, and carefully reviewed the district evaluations -- all in the locked evaluation room at Caltrans headquarters.

In the second headquarters meeting, oral presentations were made by each of the proposing consortia.

In the third headquarters meeting, the headquarters group began the process of ranking the proposals according to the evaluation criteria in the Guidelines for Conceptual Proposals. After three hours, the five-member committee decided to take a break, and begin again the next day. Matt, still in the room, stares at his copy of the results of straw polls taken by the Evaluation Committee on the SR 91 and SR 57/SAVE projects. These straw polls are reproduced in the two tables immediately below (Tables 4-4 and 4-5).

The results of the straw poll are bewildering. The Evaluation Committee tentatively ranked the SR 91 project and the SAVE project (SR57) in the third and fourth spots. All of the top four projects identified by the straw poll were located in Southern California. Based upon the statutory requirements of AB 680 and the text of the Guidelines issued by Caltrans, one of these two projects will not be approved by Caltrans, unless something changes between now and the end of the evaluation process.

Matt reviews his notes from the Evaluation Committee's first meeting, in which Robert D. Hirsch, Counsel for Caltrans reminded everyone on the committee that the evaluation criteria contained in the RFP were to be applied by members of the Committee without bias for or against specific proposals. In order to avoid potential claims of unfairness in the evaluation process by the proposers, Caltrans' counsel advised each member of the Evaluation Committee to discuss the proposals only with other members of the Evaluation Committee and only in the presence of the full Committee.

Evaluation Factor	Possible	Matt	David	Maia	Roger
Transportation service provided as a result of the proposal.	20	15	12	11	18
Degree to which proposal encourages economic prosperity and makes overall good business sense.	10	10	5	8	6
Degree of local support for proposal.	15	15	10	12	9
Relative ease of proposal implementation.	15	15	10	8	9
Relative experience and expertise of the proposal sponsors and their support team on similar projects.	15	15	15	15	14
Degree to which the proposal supports the State's environment quality and energy conservation goals.	10	6	4	5	8
Degree to which non-toll revenues support proposal costs.	5	2	1	3	3
Degree of technical innovation associated with the proposal	10	8	5	7	6
Degree of proposal's support for achieving the civil rights objectives of the State regarding the utilization of Minority and Women Business Enterprise.	10	8	5	4	7
Totals	110	94	67	73	80

Table 4-4   *SR 91 Straw Poll*

Evaluation Factor	Poss.	Matt	David	Maia	Roger
Transportation service provided as a result of the proposal.	**20**	20	13	14	17
Degree to which proposal encourages economic prosperity and makes overall good business sense.	**10**	10	7	6	5
Degree of local support for proposal.	**15**	15	9	13	12
Relative ease of proposal implementation.	**15**	15	5	7	9
Relative experience and expertise of the proposal sponsors and their support team on similar projects.	**15**	15	15	14	15
Degree to which the proposal supports the State's environment quality and energy conservation goals.	**10**	10	5	7	9
Degree to which non-toll revenues support proposal costs.	**5**	5	2	3	2
Degree of technical innovation associated with the proposal.	**10**	10	6	8	7
Degree of proposal's support for achieving the civil rights objectives of the State regarding the utilization of Minority and Women Business Enterprise.	**10**	8	6	3	5
Totals	**110**	108	68	75	81

Table 4-5   SR 57 (SAVE) Straw Poll

Matt also reviews the financial analyses prepared for the evaluation committee by Caltrans' Financial Analysis group for both SR 57 and SR 91. These are attached on the companion CD as Exhibits 4-10 and 4-11, respectively.

Exhibit 4-10   SR 57 Financial Analysis
Exhibit 4-11   SR 91 Financial Analysis

Matt remembers Hirsch's admonition about sticking to the evaluation criteria. Hirsch described their obligation this way. "The purpose of the Evaluation Criteria is to create a level playing field on which competitors for the top four projects can rely in preparing and submitting their bids. Each team probably spent at least $600,000 on its proposal, and two of the teams are reported in the press to have spent in excess of $2,000,000 in out-of-pocket expenses in an effort to win the AB 680 competition. The evaluation criteria are in effect a promise by Caltrans as to how we will evaluate the proposals. If we don't apply the pre-published criteria, the losing proposers may be able to successfully challenge the projects selected by the Evaluation Committee through a bid protest filed in Court."

The task of evaluating the proposals, ranking them, and selecting the top four seemed like an easy one at first. Early conversations with the other members of the Evaluation team seemed to confirm that their selections would be pretty easy to make. Members of the selection committee were all "career" Caltrans officials from different Caltrans departments -- David (Finance), Maia (Environmental Permitting), Roger (Human Resources), and Angela (Highway Operations). Matt had met David and Angela before, and had a great deal of respect for each of them. Maia and Roger seemed highly qualified and motivated. As the head of the evaluation group, Caltrans had arranged for John Hay, a retired director of the California Chamber of Commerce, to oversee the process and write an evaluation at the end.

Yet, the straw poll results changed all of that. It was now apparent that he and his fellow Committee members were looking at the eight proposals in different ways.

Matt recalled another bit of advice from Hirsch. "If the Evaluation Committee applies the criteria fairly to all proposals, the conclusions the Committee reaches will be presumed to be reasonable." In other words, the winners and losers chosen by Caltrans will be final.There are at least two more full meetings of the evaluation committee. Matt is very worried about what the committee will do next, and how it will be perceived both within Caltrans, by the proposers, and by numerous others public authorities that are watching the AB 680 process with unusual interest.

Questions

1. Scope of work was not defined by the State in this procurement. Both the content of the work and the qualifications of the proposers to

perform the work were in flux. What, if any, effect did this have on the "competition"?

    a. Did the evaluation criteria create a level playing field on which proposers can compete for these franchises?

    b. Was there competition among the proposers for the four AB 680 projects? Over what? Do the evaluation criteria provide a metric by which the proposals can be comparatively measured?

    c. On a comparative basis (SR 57 v. SR 91), how do the financial analyses in Exhibits 4-10 and 4-11 assist Committee members in applying the Evaluation Criteria to these two projects?

2. How should Matt prepare for the next two meetings of the Evaluation Committee?

    a. Is the taking of recorded straw polls a good idea? Why or why not?

    b. What should Matt's role be in the remaining two meetings of the Evaluation Committee?

3. The evaluation system is an unusual one. Isn't each and every possible ranking of the proposals by the headquarters review team, per se, a successful one? Is the process of selection paramount here, rather than the content of the proposals? Leave aside questions of improper influence by proposers or political forces. There is no evidence of such conduct in the case of AB 680.

4. Assume you are the CEO of one of the losing consortiums. You have been asked by your Board of Directors for a one-half page memorandum containing your evaluation as to whether the $1Million in proposal costs your firm incurred were well spent. Discuss the need for scope and evaluation criteria in this memorandum, and describe what lessons your company should learn before the next procurement similar to AB680.

References

Miller, J. B. Principles of Public and Private Infrastructure Development. Kluwer Academic Publishers, ISBN 0-7923-7201 (8), 2000.

Calleo, David P. The Bankrupting of America: How the Federal Budget is Impoverishing the Nation. First Edition. New York: William Morrow and Company, Inc., 1992.

Gomez-Ibanez, Jose A. and John R. Meyer. Private Toll Roads in the United States: The Early experience of Virginia and California. Final Report for the US Department of Transportation, University Transportation Center Region One, December, 1991.

Poole, Jr., Robert W. Private Tollways: Resolving Gridlock in Southern California. Policy Insight No. 111, The Reason Foundation. Los Angeles, 1988.

Assembly Bill 680 and Backgrounder: Enabling Legislation, A.B. 680 Summary.

Request for Qualifications (for Privatization Demonstration Project) issued by Caltrans, November 15, 1989.

Request for Qualifications for Financial Consultant issued by Caltrans, March 1990.

Guidelines for Conceptual Proposals issued by Caltrans, March 1990.

Development Franchise Agreement between Caltrans and CPTC **(Assumed in This Case to be a Draft)**, effective as of June 30, 1993.

Reinhardt, William G. "SR 91 Express Lanes Special Report: Kiewit Moves R 91 Financing to Closure, Launching a New Era in U.S. Toll Roads." Public Works Financing July/August 1993.

Regan, Edward J. "Estimating Traffic and Revenue on SR 91." Wilbur Smith Associates, Public Works Financing July/August, 1993.

Reinhardt, William G. *"Express Lanes Opened."* Public Works Financing December 1995.

Williams, Carl B. "Hot Lanes, Road Pricing and HOV Doubts, 91 Express Lanes Suggest New Directions in Highway Policy," Public Works Financing December 1996.

Notes

[14] Poole, Jr., Robert W., *Private Tollways: Resolving Gridlock in Southern California*, Policy Insight No. 111, The Reason Foundation, Los Angeles, 1988.

[15] Calleo, David P. *The Bankrupting of America: How the Federal Budget is Impoverishing the Nation.* First Edition. New York: William Morrow and Company, Inc., 1992.

[16] Assembly Bill 680 and Backgrounder: Enabling Legislation, A.B. 680 Summary

[17] Assembly Bill 680 and Backgrounder: Enabling Legislation, A.B. 680 Summary

[18] Miller, J. B. (1995). *Aligning Infrastructure Development Strategy to Meet Current Public Needs*, Doctoral Thesis, Massachusetts Institute of Technology, Cambridge.

[19] Gomez-Ibanez, Jose A. and John R. Meyer, *Private Toll Roads in the United States: The Early experience of Virginia and California*, Final Report for the US Department of Transportation, University Transportation Center Region One, December, 1991.

[20] Reinhardt, William G., *SR 91 Express Lanes Special Report: Kiewit Moves R 91 Financing to Closure, Launching a New Era in U.S. Toll Roads*, Public Works Financing, July/August 1993.
Regan, Edward J., *Estimating Traffic and Revenue on SR 91*, Wilbur Smith Associates, Public Works Financing, July/August, 1993.

Reinhardt, William G., *Express Lanes Opened*, Public Works Financing, December, 1995.

[21] Development Franchise Agreement between Caltrans and CPTC **(Treated in This Case as a Draft)**, effective as of June 30, 1993

CHAPTER 5   SANTA ANA VIADUCT EXPRESS

Infrastructure Development Systems IDS-97-T-011

Research Assistant Om P. Agarwal prepared this case under the supervision of Professor John B. Miller as the basis for class discussion, and not to illustrate either effective or ineffective handling of infrastructure development related issues. Data presented in the case has been altered to preserve confidentiality. The assistance of T. Wallace Hawkes, Executive Vice President of URS Greiner, Inc., in the preparation of this case is gratefully acknowledged.

The Problem

In the late 1980's, the California Department of Transportation (Caltrans) faced a problem common to Departments across America: how to pay for a growing list of urgent highway capital projects, with resources dwindling and maintenance costs growing. The California legislature enacted Assembly Bill 680 to authorize Caltrans to solicit proposals and enter into long-term agreements with private entities for the design, construction, lease, and operation of up to four public transportation projects; one project each was required in the northern and southern sections of the state. The purpose was to "demonstrate" that Design-Build-Operate franchises were a workable alternative to the traditional

Beginning with the enactment of the Interstate Highway System legislation in 1956, highway construction and maintenance has been financed through a combination of dedicated gasoline taxes, motor vehicle registration fees, and direct federal aid. While this combination had proved effective for over a quarter century, the anti-tax movement, particularly in California, had made raising additional taxes very difficult. The need for improving, rehabilitating, maintaining, and, indeed, expanding California's road system continued.

In the late eighties, a variety of possible methods for sustainable financing of road projects in California were being debated, including additional bond authorizations, the creation and levy of so-called "impact fees" on real estate developers, and sales taxes for transport improvements. In 1988, Bob Poole, of the Reason Foundation, proposed that the private sector might build private toll roads to fill the gap between the level of services demanded by the public and the public's apparent unwillingness to permit the government to do so directly through additional tax or use charges.

Following a conference in August, 1988, at which a number of developers made presentations describing the potential benefits that alternate delivery methods might provide to meet the State's unmet transportation needs, a number of California groups, with the cooperation of Caltrans, its then Director Robert Best, and its then Assistant Director Carl Williams successfully encouraged the California legislature, in Assembly Bill 680, to authorize Caltrans to solicit proposals and enter agreements with private entities for the construction, lease, and operation of up to four public transportation demonstration projects.

Salient Features of AB 680

AB 680 is described in the SR 91 Case presented in Chapter 4, see that description for a review of the legislation.

CALTRANS' Pre-qualification Process

A 'Privatization Advisory Steering Committee' was set up under Carl Williams and after a first stage of screening, ten potential developers were pre-qualified by Caltrans to submit 'Conceptual Proposals'. The Pre-qualification process is also described in the SR 91 Case.

Nine of the ten proposers were teams (special purpose joint ventures) of firms with different backgrounds and skills, but which together, provided the transportation, technology, design, construction, financing, and government experience Caltrans sought in its Request for Qualifications.

These joint ventures assembled an entirely different set of skills to be applied to the state's transportation needs.

The Call for Competitive Conceptual Proposals

In the second stage of Caltrans' procurements under AB 680, 'Conceptual Proposals' were requested from each joint venture. The proposers, not Caltrans, made the choice of transportation projects. The idea was to minimize expenditures by the private and public sector on the details of the particular projects prior to the time that actual proposals were selected for implementation, yet to provide sufficient information to permit a fair comparison and evaluation of competing proposals. Caltrans issued a set of guidelines for submission of conceptual proposals in March 1990.

Among the terms and conditions established in the guidelines was that each proposed project must be implemented at no cost to the State. These guidelines established weighted selection criteria in nine categories, which are also described in the SR 91 Case Study.

The Proposals

Eight proposals were received in response to Caltrans Call, including one for the Santa Ana Viaduct Express ("SAVE"), submitted by the consortium led by the Ross Perot Group. The SAVE project was evaluated by Caltrans to be the highest ranked proposal. With the highest rank, it was selected as one of the initial four AB680 demonstration projects.

The Santa Ana Viaduct Express (SAVE)

The proposal offered to provide the express tolled link shown in Figure 5-1. The proposed roadway was to connect existing State Route 57 (SR 57), Inter-State 5 (I-5) and SR 22, in the north, with I-405 and SR 73 in the south. This project was intended to complete the central regional transportation system of Orange County and to relieve congestion on the area wide roadway system.

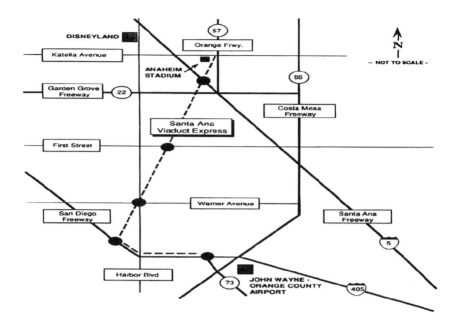

Figure 5-1   Regional Transportation Map
Figures 5-1 and 5-2 courtesy of URS Corporation.

The existing routes connecting the northern and southern parts of Orange County are very highly congested. For example, the most direct current route from the Disneyland / Anaheim Stadium area to John Wayne Airport is to travel I-5 east to SR55. These roadways are carrying 200,000 to 235,000 vehicle trips per day, taking about one hour to travel approximately 10 miles during peak hours. The I-5/SR 55/SR 22 interchange, commonly known as the 'Orange Crush', is perhaps the most congested set of roadways in all of Southern California. I-405, the major alternate route, carries over 270,000 vehicles per day in this area.

The SAVE project will add another alternative north-south route that relieves congestion, reduces travel distances, and reduces emissions in the south coast air basin through increased operating speeds. The project includes exclusive toll collection lanes, special rates for high occupancy vehicles, and uses congestion pricing concepts to improve traffic performance in the region.

The SAVE Consortium

The proposal was made by a consortium comprising a number of firms, each with expertise in fields relevant to highway construction and toll road management, including:

The Perot Group - a real estate development and investment company;

Greiner Engineering Inc. - an international, US based, bridge design and highway engineering consulting company;

The First Boston Corporation - an international investment bank;

Traffic Consultants Inc. - a consultant engaged in the planning and development of transportation infrastructure;

Amtech Systems Corporation - a designer, manufacturer, and operator of hardware and software products involving radio frequency electronic identification;

URS/Coverdale & Colpitts - consulting engineers with extensive experience in feasibility studies for toll road projects;

Robert G. Neely - the then Executive Director of the Texas High-Speed Rail Authority;

Keiwit Pacific Company - a national, US based construction company with extensive transportation experience;

Nossaman, Gunther, Knox & Elliot - a California based law firm; and

Putnam, Hayes and Bartlett Inc. - an economic and management consulting firm.[22]

Project Details

Scope

SAVE was proposed to be located in a fully developed area. Acquisition of any right of way in the region would be both difficult and expensive. To minimize acquisition costs, the Consortium advanced the novel idea of using the existing Santa Ana River flood control channel as the right of way for 8.3 miles of the 11.7 mile project. SAVE's suggestion was to build the new SR57 on a viaduct constructed at the center of this channel. Figure 5-2 shows a typical cross-section for the project within the Santa Ana flood control channel. Twin elevated structures with a width of 41.5 feet each would carry two lanes of traffic in either direction. Inter-changes with important major connecting freeways and arterial streets, such as SR 73, I-405, Warner Avenue, Harbor Boulevard, First Street, IS 5 and SR 57 were included. A 34 foot transit envelope between north and southbound lanes was reserved for future use by a transit system. The channel is part of the US Army Corps of Engineers flood protection system.

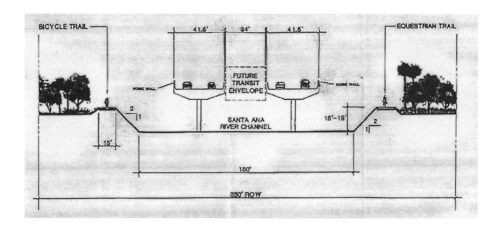

Figure 5-2   Typical Highway Section

Of the 3.4 miles of the SAVE project not located in the flood control channel, the consortium proposed that three miles be located along I-405 in the existing right of way, with the balance 0.4 miles placed alongside existing Route 73. Right of way requirements and costs are thus minimal. Toll plazas were proposed to be located at major intersections, equipped with both AVI technology and manual operated cash collection lanes.[23]

Cost Estimates

The total construction cost estimated by the SAVE consortium in July 1990 was $701,739,000, as detailed in Table 5-1 below.

COST ESTIMATES - JULY 1990	
Design and Construction	
MAINLINE	$296,026,000
CONNECTION TO EXISTING ROUTE 57	$50,355,000
I 5 CONNECTION	$28,106,000
FIRST STREET RAMP	$18,480,000
WARNER ST./HARBOR BLVD. RAMP	$14,799,000
I 405 STACKED CONFIGURATION	$88,973,000
ROUTE 73 TIE - IN	$38,195,000
MISCELLANEOUS ITEMS	$45,676,000
SUB TOTAL	$580,610,000
ADDITIONAL FEATURES:	
PAVE BOTTOM OF CorpsEng CHANNEL	$2,040,000
SCOUR PROTECTION IN CorpsEng CHANNEL	$265,000
SLOPE PAVING IN CorpsEng CHANNEL	$1,500,000
DIKE EMBANKMENT FOR CorpsEng CHANNEL	$700,000
PIER NOSES IN CorpsEng CHANNEL	$7,152,000
SUB TOTAL	$11,657,000
SUBTOTAL CONSTRUCTION ESTIMATE	$592,267,000
CONTINGENCIES	$92,356,000
CONSTRUCTION MANAGEMENT	$17,116,000
TOTAL	**$701,739,000**

Table 5-1   *Construction Cost Estimates, July, 1990*

Traffic Studies

The SR57 corridor that was the subject of the Greiner/Perot proposal had been studied by the State for a number of years, both prior to and after the passage of AB680, to estimate likely traffic on SR57 in the year 2010. The Greiner/Perot group projected volumes of 82,000, 88,100, and 101,900 vehicles per day on three key segments of the project. These estimates were reached through the following process:

- Caltrans first carried out a study to estimate the likely traffic on the project facilities using 1987 traffic levels as a base and presuming no capacity constraint on the new expressway.

- Caltrans then worked out the likely increase in traffic on this facility in 2010.

- Caltrans then adjusted these figures downward by taking a four lane capacity constraint into account.

- These estimates were then further adjusted downward by taking the effect of tolls into account.

The resulting projections of hourly average weekday traffic are given below in Tables 5-2 and 5-3.[24]

Time	Traffic	Time	Traffic	Time	Traffic
0	1				
1	.7	9	5.75	17	5.75
2	.5	10	5.25	18	5
3	.55	11	5.25	19	4.25
4	1.25	12	5	20	3.25
5	2.5	13	5	21	2.75
6	5.25	14	5.75	22	2.5
7	6	15	7	23	1.5
8	6	16	6.25	24	1

Table 5-2   Average Hourly Traffic-Northbound

Projected Toll Revenues

To finance the project, the SAVE proposal includes comparatively high tolls during peak hours and lower rates off peak -- namely, congestion pricing. The consortium projected toll revenues at $41 million to $43 million per year under the 1990 traffic conditions. Toll revenues are projected to increase to levels between $76 million and $119 million per year (in 1990 dollars) by 1997, based on tolls ranging from a peak of $ 5 per vehicle to a minimum of $ 0.25 per vehicle (in 1990 dollars). Tables 5-4 and 5-5 list proposed toll rates for different times of the day. While several rates have been proposed for AVI collection, depending on the time of the day, only two rates were proposed for cash collection. At these toll levels, volumes that are projected to divert to the SAVE facility in 1997 are 56,750 per weekday.[25]

Time	Traffic	Time	Traffic	Time	Traffic
0	1				
1	.7	9	5.5	17	5.5
2	.5	10	5.25	18	5.25
3	.55	11	5.25	19	4.5
4	1	12	5.5	20	3.75
5	3.75	13	6	21	3
6	7	14	5.5	22	2.25
7	7.5	15	6	23	1.25
8	6.75	16	6	24	

Table 5-3   Average Hourly Traffic-Southbound

SAVE'S Financing Plan

The consortium's financing plan was separated into three phases - Pre-construction, Construction, and 'Take-Out'.

Pre-Construction

Programmed to occur between January 1990 and July 1993, pre-construction was estimated to cost $47 million, and was further subdivided into four sub-phases:

NORTHBOUND				
	1990		1997	
TIME	AVI	CASH	AVI	CASH
MIDNIGHT	$0.25	$0.25	$0.25	$3.00
1:00 AM	$0.25	$0.25	$0.25	$3.00
2:00 AM	$0.25	$0.25	$0.25	$3.00
3:00 AM	$0.25	$0.25	$0.25	$3.00
4:00 AM	$0.25	$0.25	$0.25	$3.00
5:00 AM	$0.75	$3.00	$1.00	$3.00
6:00 AM	$2.75	$3.00	$5.00	$5.00
7:00 AM	$5.00	$5.00	$5.00	$5.00
8:00 AM	$5.00	$5.00	$5.00	$5.00
9:00 AM	$4.50	$5.00	$5.00	$5.00
10:00 AM	$3.75	$3.00	$5.00	$5.00
11:00 AM	$3.75	$3.00	$5.00	$5.00
12 NOON	$3.25	$3.00	$5.00	$5.00
1:00 PM	$3.75	$3.00	$5.00	$5.00
2:00 PM	$5.00	$5.00	$5.00	$5.00
3:00 PM	$5.00	$5.00	$5.00	$5.00
4:00 PM	$5.00	$5.00	$5.00	$5.00
5:00 PM	$5.00	$5.00	$5.00	$5.00
6:00 PM	$3.75	$3.00	$5.00	$5.00
7:00 PM	$1.00	$3.00	$1.25	$3.00
8:00 PM	$0.75	$3.00	$1.00	$3.00
9:00 PM	$0.75	$3.00	$0.75	$3.00
10:00 PM	$0.50	$0.25	$0.25	$3.00
11:00 PM	$0.25	$0.25	$0.25	$3.00
AVERAGE	$2.53	$2.78	$2.95	$4.08

Table 5-4   *Illustrative Time of Day Tolls-Northbound*

SOUTHBOUND				
	1990		1997	
TIME	AVI	CASH	AVI	CASH
MIDNIGHT	$0.25	$0.25	$0.25	$3.00
1:00 AM	$0.25	$0.25	$0.25	$3.00
2:00 AM	$0.25	$0.25	$0.25	$3.00
3:00 AM	$0.25	$0.25	$0.25	$3.00
4:00 AM	$0.25	$3.00	$0.25	$3.00
5:00 AM	$0.75	$3.00	$1.00	$3.00
6:00 AM	$5.00	$5.00	$5.00	$5.00
7:00 AM	$5.00	$5.00	$5.00	$5.00
8:00 AM	$5.00	$5.00	$5.00	$5.00
9:00 AM	$2.75	$3.00	$5.00	$5.00
10:00 AM	$2.25	$3.00	$5.00	$5.00
11:00 AM	$2.25	$3.00	$5.00	$5.00
12 NOON	$2.75	$3.00	$5.00	$5.00
1:00 PM	$3.25	$3.00	$5.00	$5.00
2:00 PM	$4.75	$5.00	$5.00	$5.00
3:00 PM	$4.50	$5.00	$5.00	$5.00
4:00 PM	$4.75	$5.00	$5.00	$5.00
5:00 PM	$3.25	$5.00	$5.00	$5.00
6:00 PM	$2.75	$3.00	$5.00	$5.00
7:00 PM	$1.00	$3.00	$1.25	$3.00
8:00 PM	$1.75	$3.00	$1.00	$3.00
9:00 PM	$0.75	$3.00	$0.75	$3.00
10:00 PM	$0.50	$3.00	$0.50	$3.00
11:00 PM	$0.25	$0.25	$0.25	$3.00
AVERAGE	$2.27	$3.01	$2.96	$4.08

Table 5-5   Illustrative Time of Day Tolls-Southbound

Proposal preparation

Approximately $1.5 million was spent on proposal preparation, contributed as equity by consortium members. Return sought on this investment in the SAVE proposal was 28 %.

Project Development

Upon selection of the SAVE project by Caltrans, the consortium's financing plan estimated that $2.3 million next phase would be spent on further development and refining of the project up to the point that a Development Franchise Agreement was executed with Caltrans. Return sought on this investment in the SAVE proposal was 28 %.

Project Initiation

Following execution of the Development Franchise Agreement, traffic and revenue projections would be validated, environment clearances obtained, and preliminary meetings with equity investors and lenders held to obtain indications of interest from private financing sources. This phase was estimated to cost $12 million in the financing plan. SAVE's proposal stated that until this phase was complete, project feasibility was uncertain. Return sought on this investment in the SAVE proposal was 28 %.

Final Pre-Construction

This stage, estimated at $31 million, focused on obtaining all administrative clearances, mobilizing for construction, and negotiating financial commitments. Return sought on this investment in the SAVE proposal was 28 %.

Construction

SAVE's financing plan assumes that construction would occur over five years between October 1992 and June 1997, and would require $741.7

million (1990 dollars). It was estimated that the Design-Build project delivery method would save one year (20%) over the typical public works construction process, in which design is segmented from construction.

Based upon SAVE's proposal, the draw down schedule for project funds was projected to be as follows:

Up to September 1993	14 %
October - December 1993	4 %
January - March 1994	6 %
April - June 1994	6 %
July - September 1994	6 %
October - December 1994	6 %
January - March 1995	8 %
April - June 1995	8 %
July - September 1995	8 %
October - December 1995	8 %
January - March 1996	6 %
April - June 1996	4 %
July - September 1996	4 %
October - December 1996	4 %
January - March 1997	4 %
April - June 1997	4 %

In an effort to reduce accumulated interest costs, several strategies were being considered by SAVE:

First, every effort would be made by the consortium, both collectively and by individual members, to accelerate the design/construction interface process, and to fast track construction, equipment delivery, and installation, wherever possible.

Second, early construction draws would be financed out of a combination of debt and equity. An equity contribution of $50 million from an independent third party investor would be used to pay 50% of construction draws, until such time as it was depleted. Return sought on this equity contribution in the SAVE proposal was 28%.

Third, debt would only be incurred as and when needed, even though up front commitment charges would be paid. Interest was assumed at 11% per year.

Take - Out

SAVE's financing plan states that between June 1997 and April 1999, after construction is completed and the revenue stream begins, the consortium would arrange for multi-tiered long term take-out financing composed of:

Senior Lien Toll Road Revenue Bonds, which would be the primary source of take out financing and would be secured by a first charge on net toll revenues;

Subordinated bonds, to be secured by a pledge of net revenues not already committed for the project's Renewals and Replacement fund, Reserve fund, or for servicing the Senior Lien Toll Road Revenue Bonds;

Long term equity, to be obtained from institutional and high net worth investors, local governments, local businesses, and consortium members; and

Ancillary Supplemental Revenue, This tier refers to the consortium's hope that, as permitted by AB680, a series of Area Development Projects and Commercial Property Developments would occur in the air space above, below, and adjacent to the project, providing further long term, and therefor, financible revenue streams.[26]

Construction Schedule

The SAVE proposal assumed a franchise start date of January 1990, with the completed facility open for traffic in April 1997. SAVE's lease was expected to last for 35 years, until April 2032. The project schedule and important milestones are set forth in Exhibit 5-1, on the companion CD along with Exhibit 5-2 which shows the financial analysis of the anticipated project results.

The Final Franchise Agreement

The final agreement allowed a base rate of return of 20.25%, the highest return allowed by Caltrans on any of the AB680 projects. This relatively high rate of return has been attributed to the risk of this project, particularly the uncertainty in projected toll revenues.

Questions

Task 1 - Financial Analysis   (Base Case)

Caltrans picked the SAVE proposal as the best of eight proposals it received. Yet, the project includes relatively high peak hour tolls and an uncertain number of paying customers.

Carry out a financial analysis and calculate the Net Present Value of the project, using the following assumptions:

a. Construction Cost $701,739,000
    Pre construction Expenses $47,000,000
    Other Miscellaneous Costs
            Right of Way $40,000,000
            Bank Annual Charge $7,000,000
            One time Commitment Charge $14,000,000
            Financial Adviser Fees $10,000,000

Pre-construction period begins in 1990 and ends in June 1993.
Construction period begins in October 1992 and ends in June 1997.

Projected schedule of expenses as a percentage of relevant costs:

Year	Pre construction	Construction
1990	0	0
1991	25	0
1992	55	2
1993	20	16
1994	0	24
1995	0	32
1996	0	18
1997	0	8

Right of Way is paid for in 1993 and the one time Bank Commitment Charge and Financial Advisers Fees are paid in 1992. Annual bank charges are payable from 1993 to 1997 (i.e. the year of completion of the project)

b. A cash subsidy of $ 15,000,000 is receivable from Orange County in 1991, to partially defray pre-construction expense

c. Lease Period 35 years from construction completion date

d. Average Traffic during first year of operations 50,000 vehicles per weekday, in each direction

e. Average Toll per vehicle - $4.00

f. Annual increase of toll revenues 10% (resulting both due to increased traffic and increased tolls)

g. Discount Rate - 10%

h. O&M Costs $17.2 million in the first year of operations and increasing at 2.0% every year

Task 2 – Sensitivity Analysis

The financial viability of such projects needs to be tested by varying the assumptions that may not be realized and to verify the robustness of the project's feasibility under different scenarios. Such a sensitivity analysis is critical for deciding on the viability of a project.

Carry out a sensitivity analysis, each against the Base Case Analysis, with the following changes in the assumptions:

a. Discount Rate - Calculate the Net Present Value using discount rates of 5% and 15%. What is the highest discount rate that the project can sustain under these assumptions, i.e., the rate at which the NPV becomes zero?

b. Construction Cost - What happens to the NPV if the construction cost goes up by 10%, 20%. What is the highest cost escalation that the project can sustain? (Assume the same schedule of expenses as for the base case).

c. Delay in completion - Delay can occur either in the pre-construction stage on account of delays in securing clearances, tying up funds, acquiring right of way, etc. or during the construction stage on account of slower progress of construction work than anticipated.

Calculate the impact of both a 3-year and a 5-year delay at the pre-construction stage. Assume that some administrative expenditure would have to be borne during this period and would have the net effect of inflating project costs. Assume that during each year of delay an expenditure of 2% of the pre-construction costs is incurred. Thereafter, the schedule of expenditure given in the base case becomes applicable. Delay occurs after the second year of pre-construction. As an example, with a five year delay, the schedule would be as follows:

a. Year 1   0%
b. Year 2   25%
c. Year 3   2%
d. Year 4   2%
e. Year 5   2%
f. Year 6   2%
g. Year 7   2%
h. Year 8   55%
i. Year 9   20%

d. Calculate the effect on NPV if the project is delayed by one year at
    the construction stage. Assume that the total project cost goes up by
    5 % as a result of this delay. Also assume that, because of this
    delay, the schedule of expenses is revised as follows:

	Year	% of Capital Cost
a.	Year 1	0
b.	Year 2	10
c.	Year 3	15
d.	Year 4	25
e.	Year 5	20
f.	Year 6	15
g.	Year 7	15

e. Toll Revenue Projections - What happens to the NPV if the toll
    revenues for the first year fall short of expectations by 10%? by
    20%? by 30%?

f. What is the effect on NPV in the original Base Case, if the Cash
    subsidy from Orange County is not available, because of an
    unanticipated general default by the County with respect to
    unrelated financial obligations.

Task 3 – Cash Flow Analysis
(Pre-construction and Construction stages)

Financing of large projects is both complicated and difficult. The main
objectives of the project financing plan are to ensure timely availability of
funds at market rates of return and to minimize the interest burden on the
project. To facilitate this, financial managers carry out a cash flow analysis
and aim to minimize or delay loan draws. Equity investments early in the
project reduce debt interest charges, and may reduce dividend payments.
The tradeoff to project promoters making equity contributions is that returns
of equity investments typically come last, after other equity and debt is
repaid, and only if the project makes a profit.

Pre-construction expenses are often contributed by the principal
promoters, since risks are highest at this stage, and both equity and debt

financing is both difficult to find and expensive, anyway. Indeed, the nature and extent of the contribution of equity by reliable project promoters sends a strong signal to investors as to the promoter's view of a project's long-term financial and technical viability.

As project details are tied up, with schedules and costs increasingly stabilized, it is progressively easier to secure debt financing at competitive interests. After project completion, the history of project revenue streams provides reliable bases for long term take-out financing.

While the financial analysis is usually carried out for a base year, cash flow statements are often prepared taking the effect of inflation and cost escalation into account. Whether cost escalations are included or not, consistent treatment is required, with a clear statement of assumptions made as to inflation and escalation.

This task requires you to carry out a cash flow analysis for the pre-construction and the construction phases of the project in order to appreciate the importance of good cash management. Do the analysis for the period from 1991 to the project completion date of June 1997.

For this analysis, the following assumptions are to be made:

a.   Costs for pre-construction and construction periods as well as other miscellaneous costs are the same as in the financial analysis of Task 1.

b.   An escalation factor of 4.5% per year on all costs, except bank charges and financial advisers fees. The analysis should be done for each quarter separately with the escalation factor equally divided for each quarter.

c.   An expenditure of $1.5 million has been made for preparation of the conceptual proposal, and appears as a negative opening balance for the cash flow analysis from 1991.

d.   Outside equity of $10.5 million is available for the pre-construction stage as per the following schedule:

1991	$2.5 million
1992	$6.0 million
1993	$2.0 million

One quarter of each of these amounts is available in each quarter of the respective year for pre-construction expenses.

e. Equity contributions from sources other than the consortium is available to be drawn to the extent of $50,000,000, beginning with the first twelve months of actual <u>construction</u>, i.e. in the 4th quarter of 1992.

f. The consortium contributes whatever equity is required to fund pre-construction expenses, less what is available from outside sources. For the construction expenses there is no equity contribution from the consortium and debt financing is resorted to on a 50% basis until the outside equity of $50 million available from outside sources is exhausted.

g. Any surplus or deficit at the end of a quarter is carried over to the next quarter.

h. Interest on all outstanding loans is at 11% and has to be paid from the quarter immediately following the quarter in which it is obtained. However, the principal is repayable only after project completion and out of debt financed proceeds.

From the cash flow analysis, answer the following questions:

1. What is the escalated cost of the project?

2. What was the total amount of the construction debt financing, and how much take-out (possibly bond) financing do the project sponsors have to raise to repay all outstanding construction related loans?

3. What is the total interest burden that accrued during the project construction period (from October 1992 to April 1997)?

4. What were the total payments that were made to the bank towards annual fees and bank commitment charges? What percentage of the total escalated project cost do these payments represent?

5. What would be the effect of the equity contribution being $ 100 million instead of $50 million on the project cost, interest burden and total loan?

Task 4 – Cash Flow Analysis - (Take out Stage)

Background

Once design and construction are complete, most uncertainties associated with the original estimate of cost and schedule have become realities, for better or worse.   The risk associated with those uncertainties that remain - for example, the volume of use, or the cost and schedule of repair, maintenance, and operations -come down considerably. This enables borrowing on better terms, either through lower interest rates or longer repayment schedules.   Furthermore, a completed project that is producing revenues permits such revenues to be used as security for long term financing. A variety of long term financing arrangements are possible, based upon the securitization of a relatively reliable revenue stream.

For example, 'Revenue Bonds', which carry a fixed interest payable every year by the Bond Issuer (borrower) to the Bond Holder (Lender) are readily traded in the market and offer considerable flexibility to bond holders. To protect lenders, such bonds are issued with a certain degree of cover, i.e., the net revenues (total revenues less O&M expenses) are not only able to pay the annual interest due but also provide an extra cushion to cover unforeseen fluctuations. Typically, the coverage is 1.3, meaning that the revenues should be able to pay 130% of the annual interest, after adjusting for O&M expenses.

Borrowers normally seek to maximize the amount they can raise by way of bonds but are constrained by the revenue flows and the need for a security cover. The target size is usually the entire amount of outstanding loan and compounded equity at the end of the construction phase plus the administrative costs of raising the bonds. Whatever cannot be raised through such bonds has to be raised through second or third levels of subordinate debt (often referred to as tranches), through contributions of equity by project promoters, and in most cases, a combination of both. Subordinate debt is debt that is serviced after higher priority debt serviced. Because of the preferential treatment of primary debt, and the relative additional risk associated with lower tiers of debt, subordinate debt typically carries an interest rate that is higher than the interest rate long term bonds.

In the SR57 proposal submitted by the Perot Group, equity proposed to be contributed at the pre-construction and construction stages was carried by the project promoters at a fixed discount rate (28%). This discount rate reflects the fact that project promoters receive a return on their initial investment, if at all, behind all other project investors.

Reserve funds are also typically established early in the project, while cash flows are being estimated and refined. One such fund in the SR57 proposal is the "Replacement & Renewals Fund". Another similar fund is the "General Reserve Fund", used as a cushion to provide debt service during unexpectedly bad years. This normally has a target balance that is a percentage of the total bond issue.

1. Carry out a cash flow analysis for the take out stage of this project, using the assumptions set forth below and covering a period of 35 years. You are required to assess the extent of bond financing to be raised and the equity contributions that are needed from project promoters.

Assumptions

The following assumptions may be made for this question:

a. Interest Rate on Bonds Issued 11%
b. Coverage Ratio 1.3
c. First Year Toll Revenues $104 million
d. Annual increases in toll revenue 10%
e. O&M Expenses in the first year of operations $17.2 million
f. Annual increase in O&M expenses 2%

g. Compound Rate for pre-construction and
h. Construction equity 28%
i. Contribution to Replacement & Renewals Fund @ 0.6% of the Project construction cost (1990 prices) and right of way costs to be contributed every year during the first seven years
j. General Reserve Fund target 10% of the bond issue
k. (contributed to the extent of surplus available after long term debt service and contribution to Replacement & Renewals fund)
l. Interest earnings on reserve funds is10%
m. Bond issue cost 2% of the targeted amount
n. Targeted amount less the bond issue amount is obtained as equity and subordinate debt. Half is subordinate debt and the balance is equity. Subordinate debt has an interest rate of 15%.

After carrying out the base case cash flow analysis above, you are required to determine the following:

2. What is the rate of return on long term equity, i.e., what is the discount rate at which the total Net Present Value of dividends earned by equity holders equals the Net Present Value of their equity contribution.

3. What is the effect on rate of return on long term equity if the bonds carry an interest of 8%, 12%?

4. What is the effect on rate of return for long term equity if the consortium had been able to raise an additional 15% by way of bonds (debt financing) compared to what was believed possible in item 1 above? What is the effect if they were only able to raise 15%less than anticipated in item 1 above?

5. What is the effect on the rate of return for long term equity in the base case, if the subordinate debt carries an interest rate of 20%?

Task 5 - Risk/Reward Allocation

Assume that the above analyses have been conducted prior to submittal of the SAVE proposal to Caltrans. These analyses have been shared among all consortium members. Each member of the team is familiar with:

a. the relationship between the project schedule and project costs;

b. how sensitive the interrelationships are between schedule, design, construction, and cost;

c. the total equity contribution required to produce the project within the schedule and budget;

d. the total debt financing required, at assumed market discount rates, to finance the balance of the project;

e. expected NPV, based upon the current best guess at project schedule, quality, and cost; and

f. rates of return required for debt financing, equity contributions from third parties, and hoped for rates of return for consortium equity contributions.

A few examples will illustrate these points. Both late and early design completion by Greiner will affect the economic performance of the consortium, as will late/early completion of construction by Kiewit Pacific. Completion of design and construction within budget will also directly affect the economic performance of the consortium. If the traffic volumes projected by URS/Coverdale are significantly misstated, with respect to either volume or time period, the revenue side of the cash flow forecasts will be directly affected. Should the electronic toll equipment supplied by Amtech Systems fail to properly identify and charge toll revenue properly and quickly, a similar impact is likely on the revenue side. If revenues are insufficient to pay debt service on or ahead of schedule, you can assume that bank guarantees signed by key consortium members will require additional equity contributions from one or more of Greiner, the Perot Group, Kiewit Pacific, and Amtech.

Next Tuesday the consortium will meet to finalize its proposal. The purpose of this meeting is to finalize the intra-consortium agreement as to how equity, contracts for services, profits, losses, bonuses will be allocated among the participants. Some of the key elements of the project are within the control of the parties, such as initial equity contributions, work assignments, and timely and proper performance of these work assignments. Other elements of the project may be beyond the ability of any party, including the government, to control.

You have been asked by the group to chair this meeting, and to suggest a framework in which equity, contracts, profits, losses, bonuses will be allocated among the group. The group agrees with your goals, which are these:

- fairly compensate each member for services rendered in connection with project development, proposal preparation, pre-construction, construction, and M&O;
- provide strong incentives for members to deliver those elements within their primary control - on or before the time scheduled, with uniformly high quality, at or better than budget;
- provide incentives for the entire team to exceed projected time, quality, and cost projections for the project (for example, a design improvement that results in M&O savings produces a financial reward to the designer and the M&O operator);
- provide an appropriate return for those consortium members contributing equity to the project.

Exhibit 5-3, on the companion CD, is a tabular description of the context in which these issues will be discussed at the meeting. Which of the parties are taking what risks over what periods of time with how much equity contributions for what potential gain are difficult questions. Sometimes, consultants that are unwilling to make equity contributions will consider performing services at cost, or at lower than typical markups, in exchange for an incentive kicker.

Exhibit 5-3   *Anticipated Project Results at Base Budget & Schedule*

1. What will you propose to the group next Tuesday? Consider an approach which rewards solid performance, early equity contributions, yet encourages savings in the project's (a) contingency fund, (b) interest expense, and (c) the generation of unexpected savings. At the end of the

meeting next Tuesday, you hope to either establish a mechanism to fill the gray cells in Exhibit 5-3 or to have some of them actually filled in by agreement.

Does your approach focus upon a formula? a compensation committee? a mixture of both?

2. Assume that both financing banks and the bond markets are looking for financial strength in the consortium. What form of organization should the consortium take? How would you propose that it be capitalized?

> Partnership?
> Joint Venture of Corporations?
> A New Corporation?

References

Assembly Bill 680.

Request for Qualifications issued by Caltrans, dated November 15, 1989.

Request for Conceptual Proposals issued by Caltrans, dated March 1990.

Request for Qualifications (Financial Consultant) dated March 1990.

Proposal for a Toll Revenue Transportation Project from the Perot Group, dated August 1990.

California "State Transportation Investment Plan" of 1989.

Calleo, David P. The Bankrupting of America: How the Federal Budget is Impoverishing the Nation. William Morrow and Company, Inc., New York, 1992.

Gomez-Ibanez, Jose A. and John R. Meyer. Private Toll Roads in the United States: The Early experience of Virginia and California. Final Report for the US Department of Transportation, University Transportation Center Region One, December, 1991.

Poole, Robert. "Private Tollways: Resolving Gridlock in Southern California." Policy Insight No. 111, Reason Foundation. May 1988.

Notes

[22] Proposal for a Toll Revenue Transportation Project from the Perot Group, dated August 1990, pp. 21 to 32

[23] Ibid., pp. 48-55

[24] Ibid, pp. 94 to 100

[25] Ibid, pp. 125

[26] Ibid, pp. 150 to 161

CHAPTER 6   CONFEDERATION BRIDGE OVER THE NORTHUMBERLAND STRAIT

Infrastructure Development Systems IDS-97-T-010

Research Assistant Om P. Agarwal prepared this case under the supervision of Professor John B. Miller as the basis for class discussion, and not to illustrate either effective or ineffective handling of infrastructure development related issues. Data presented in the case has been altered to preserve confidentiality. The assistance of Jim A. Feltham, P. Eng., Public Works and Government Services Canada, Paul Giannelia, of SCI, and Krista Jenkins of SCDI in the preparation of this case is gratefully acknowledged.

Introduction

Prince Edward Island (PEI), with a population of 129,765 (June 1991) and an area of 5660.38 sq. km., lies off the east coast of Canada, being separated from Nova Scotia and New Brunswick by the Northumberland Strait. It is one of the four Atlantic Provinces of Canada and measures 224 km. from tip to tip with a width ranging from 6 to 64 km.

Principal Economic Features of Prince Edward Island

Agriculture and fishing dominate goods production and food processing dominates manufacturing. Tourism is an important contributor to the economy. The provincial GDP in 1993 was C$ 2,349 Million having increased from C$ 1,924 Million in 1989.

Potatoes are a major source of farm income, contributing an average of more than 30 % of the total farm receipts. Annual farm receipts exceed C$ 222 Million. There are approximately 5,000 people employed in agriculture and out of 640,000 acres devoted to agriculture, 381,000 are under crops.

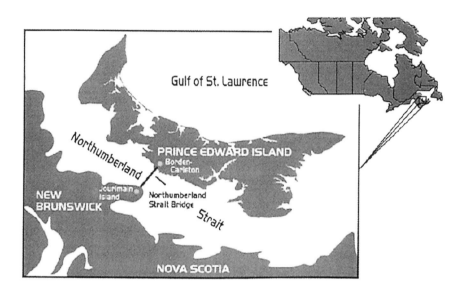

Figure 6-1   Regional Map Showing Project Location

Fishing and aquaculture are quite important contributing in excess of C$ 210 Million annually. The lobster fishery accounts for about two-thirds to three-quarters of the annual fishing income. There are approximately 5,300 fishermen and another 2,500 persons are employed in the fish processing industry.

A large part of the island's manufacturing sector is involved in the processing of agricultural and fisheries products. Specialized manufacturing industries have also been established for producing such goods as diagnostic medical kits, optical frames and steel and aluminum cookware.

There are some 280,000 hectares of forested land on PEI. In 1993, 36 million board feet of lumber, worth C$ 14 Million was cut on the islands saw mills. Pulpwood sales to markets in Nova Scotia, New Brunswick and Newfoundland are also sizable.

Tourism is an extremely important industry to the economy. In 1993 tourist spending amounted to C$ 121 Million. There were approximately

690,000 visitors to PEI during that year. Table 6-1 shows the average annual (all items) Consumer Price Index (base = 1986) for prior years.

Year	For Canada		For PEI	
	Annual Aver. CPI	% Δ	Annual Aver. CPI	% Δ
1989	114.0	-	111.5	-
1990	119.5	4.8	117.1	5.0
1991	126.2	5.6	125.9	7.5
1992	128.1	1.5	126.9	0.8
1993	130.4	1.8	129.3	1.9
1994	130.7	0.2	129.0	- 0.2

Table 6-1   Consumer Price Index for Canada and Prince Edward Island
Source: 21st Annual Statistical Review of Prince Edward Island

Transport Links with the Mainland

PEI became a part of the Canadian confederation in 1873. Under the terms of its entry into the confederation, the Government of Canada agreed to provide a continuous and efficient year round transportation facility for goods, services and people between the island and the mainland. The Canadian Government has fulfilled this responsibility by operating subsidized ferry services between the island and two points on the mainland. A year round service from the island to Cape Tormentine in New Brunswick is run by Marine Atlantic, a crown corporation. This service operates every hour from 6:30 am to 1:30 am, with a slightly reduced frequency during weekends. The crossing takes about 45 minutes. This service is supplemented by another to Caribou in Nova Scotia from May to December (the ice-free period), operated by Northumberland Ferries Ltd., a private company.

Traffic Carried and Toll Rates

The traffic carried by the two ferry services from 1992 to 1994 is shown in Table 6-2. Over the years there has been a modest annual growth of around 3 to 4% in the traffic carried.

Year	1992	1993	1994
PEI-Cape Tormentine ferry			
Autos & Pickup trucks	600,911	621,577	638,017
Recreation Vehicles	25,053	27,104	28,596
Buses	3,231	3,197	2,905
Motorcycles & Bicycles	4,729	4,670	4,638
Commercial Vehicles	148,626	159,581	187,986
Passengers	1,704,022	1,772,219	1,820,332
PEI-Caribou ferry			
Autos & Pickup trucks	167,102	158,545	185,753
Recreation Vehicles	9,431	9,625	11,749
Buses	1,220	1,117	1,083
Motorcycles & Bicycles	3,133	2,981	3,481
Commercial Vehicles	22,673	24,339	27,678
Passengers	499,798	489,488	584,187

Table 6-2   *Ferry Service Traffic from 1992-1994*
Source: *21st Annual Statistical Review of Prince Edward Island*

Fares are charged for the ferry only while leaving PEI and at the following rates:

Each adult passenger	C$ 8.00
Children between 5-12 years	C$ 4.00
Children below 5 years	Free
Each Vehicle	C$ 19.50
Vehicles with trailers of 20 - 30 feet	C$ 29.25
Vehicles with trailers of 30 - 40 feet	C$ 41.50
Vehicles with trailers of 40 - 50 feet	C$ 53.75

Difficulties with Ferry Services

The ferry services have produced complaints of delays at terminals, particularly during the summer months when PEI is an attractive tourist destination. PEI's economy is commonly believed to have suffered from the absence of a fixed link, although competing arguments have been advanced as to the long-term value of the relative isolation of the island that the absence of such a link promotes. Operation of the ferries has been an expensive proposition for the government as well with the future projections being of increased government spending on the aging fleet, which requires repairs and modernization. The losses in ferry operations, after taking into account the revenues from fare collections is required to be subsidized by the Government. In 1992 the Government spent approximately C$ 42 Million subsidizing these ferry operations.   These subsidies rise at a rate approximately 15 - 20 % higher than the Consumer Price Index (CPI). Projections made in 1987 of future operating and capital costs are given in Exhibit 6-1, on the companion CD. The magnitude of future expenses and continuing subsidies is of concern to the national Government.

Exhibit 6-1   1987 Projections of Ferry Operations and Capital Costs

Proposal for a Fixed Crossing

Whether the ferries have proved to be cost effective and efficient has been discussed for a number of years. Many have advocated replacement with a more efficient fixed crossing. Unsolicited proposals have been received over a number of years to construct such a connection, including proposals for a tunnel, and a combined causeway-tunnel-bridge.

The Government of Canada decided to take up a fixed crossing using what it described as a Build-Operate-Transfer (BOT) approach. It invited the private sector to design, construct, finance, and operate a fixed crossing in exchange for a commitment by the government to redirect the funds currently spent on ferries to the long-term finance of a fixed crossing. Stage I of the invitation process, completed in 1987, resulted in the selection of seven proposers who demonstrated the potential to proceed further. Six proposed bridges and one proposed a tunnel. In Stage II, an analysis of the technical proposals against stated evaluation criteria resulted in the shortlisting of three of the proposal teams. The technical criteria were primarily aimed at assessing the ability of the developer to design, construct, operate and maintain the fixed crossing. All the shortlisted proposals were for bridges. Stage III of the procurement process involved an evaluation of the financial proposals. Ultimately, this process led to the selection of Strait Crossing Development Inc. (SCDI) as the chosen franchisee/developer.

The final agreement was delayed on account due to concerns raised regarding the environmental effects of the bridge. After an extensive public review process, a number of additional elements were added to the project. The final franchise documents were signed in October 1993.

Terms of the Franchise Agreement

Per the agreement, the bridge is to be designed, built, financed, operated, and maintained by SCDI for a lease period of 35 years, commencing from June 1997. The agreement promises that ferry operations will cease and SCDI will be entitled to collect a toll from users of the bridge. In the first year, these tolls may not exceed those charged by the ferry services.

Thereafter, the tolls may be escalated annually at a rate not exceeding 75 % of the annual increase in the CPI. The agreement also provides that SCDI will receive an annual subsidy equivalent to C$ 41.9 (in 1992 dollars) from the Government of Canada. The first installment of this subsidy will be paid at the end of May 1997, whether the bridge is completed or not. If the bridge is not complete by that time, ferry services will continue to operate at SCDI's expense.

On expiry of the 35 year lease period, ownership of the bridge will automatically transfer to the Government of Canada for a nominal consideration of C$ 1.00.

Cost Estimates and Project Schedule

The estimated cost of the bridge is C$ 840 Million and SCDI is financing this through a combination of debt and equity. Preliminary work on the project commenced in early 1993 and the bridge is scheduled for completion by May 1997. See Exhibit 6-2 for the project schedule.

Exhibit 6-2   Project Schedule

Bridge Design and Construction Plan

The proposed fixed crossing is to run from just west of the ferry terminal at Cape Tormentine to a point adjacent to the terminal on PEI. It is a shore to shore bridge of a total length of 12.9 km. The main bridge, over deeper waters, has a length of 11 km comprising 44 spans of 250 m each. Pre-cast concrete balanced cantilever spans alternate with drop-in spans. There are also two approach bridges on either side of the main bridge, connecting it to the shore and designed for comparatively shallow waters. The approach bridge on the PEI side has a length of 580 m and comprises of 7 spans. The one on the Cape Tormentine side has a length of 1300 m and has 14 spans. The typical span length of the approach bridges is 93 m. The approach bridges are pre-cast segmental balanced cantilevers. Various design specifications of the bridge are given in Exhibit 6-3.

Exhibit 6-3   Bridge Specifications

Of specific concern were the effect of ice on the structure and the subsequent effects of this on the ice climate on the strait, including navigation. To minimize the adverse effects of ice formation near the bridge, conical ice shields have been provided on the piers at water level. Exhibit 6-4 shows a cross section of the pier structure with the conical structure. To minimize disruption to the floor of the strait, and to minimize ice blockage of the strait, piers were spaced at 250 m intervals. The design requirement is that the addition of the bridge not delay the breakup of ice in the straight for more than two days once in a hundred years.

Exhibit 6-4   Cross Section of Typical Pier
By Permission, ©Public Works and Government Services Canada

On account of the adverse weather conditions at sea during winter months, erection work is possible only for 34 weeks in a year. The construction philosophy was to do most of the work on land. Apart from reducing the dependency associated with marine work, this also maximizes the use of local labor skills and minimizes disturbance to the marine environment. Construction of the bridge components takes place on land, followed by transport and assembly in place on the sea.

Two hydraulic sleds were used to transport concrete components. In addition, eight tower cranes and four gantries were also employed in moving the heavy components. Elaborate yards were built at both terminals for pre-casting the bridge components. The yard on the PEI side was built like an assembly plant in which components move between fixed formwork stations. Timing of the pre-casting operation depends on the marine erection schedules.

The first step in casting cantilevered spans was to cast their centers, which are 14.7 m deep and 17 m long. Each contains an inverted 'V' steel diaphragm. These double as frames for the concrete encasement and are 13.6 m tall and 10.2 m wide at the base. They are plant fabricated in four pieces and field bolted.

At the next formwork station, a 7.5 m long segment was added to each side and this process was repeated through seven more stations where additional lengths were added. Each component was cured for 30 hours before being stressed.

A self-propelled floating crane, the 'Svanen', carried components to the bridge site. Members 52m and 60m function as simply supported long drop-in spans. The drop-ins were cast in a 50m long form in one continuous pour. Formwork was then installed at the span ends to cast either a diaphragm for a continuous connection or a corbel for an expansion joint.

In the casting phase of the approach bridges, each segment was matched to the adjoining segment through a match-casting process. This process involved casting a segment and then casting the next segment against a matching face of the previous section. The segments are made of reinforced concrete, post tensioned in the transverse direction. The segments were erected in a balanced cantilever fashion using longitudinal post-tensioning steel tendons to permanently join the segments.

The main bridge components were produced at a staging facility on PEI. There were four separate component production areas to produce the main girders, drop in spans, pier shafts, and pier bases. The components were made of reinforced concrete, post tensioned where required.

The erection sequence included verification of the geological condition of the pier foundations, excavation of the pier locations, placement of positioning pads on the excavated location, placement of the pier base, underwater concrete work at the pier base to ensure 100% contact with the bedrock, placement and post-tensioning of the pier shaft, placement of main girder and post-tensioning to the shaft, placement of the drop-in spans between main completed main spans, continuous post-tensioning of structural frames and grouping of expansion joint bearings between frames.

Exhibit 6-5, on the companion CD, includes a number of pictures of the different bridge components and construction work in progress.

Exhibit 6-5   Construction Progress Photos
By Permission, ©Public Works and Government Services Canada. Credit Boily Photo, Provincial Airlines Ltd., Ron MacDougall.

The Developer's Consortium

The successful proposer essentially built and operated an outdoor manufacturing plant to mass-produce the four key components of the bridge. The strategy was conceived prior to submittal of the proposers, which allowed the government to share in the time and cost savings that were achieved through strong and early planning. The presentation of the project as a DBO/BFO opportunity allowed the integration of design, construction and operations to take place.

At the time of franchise award, the developer's consortium included the following:

Strait Crossing Development Inc. is a Canadian consortium made up of Northern Construction, GTMI (Canada) Inc., Strait Crossing Inc., and Ballast Nedam Canada Ltd.

Strait Crossing Inc. is a 100 % Canadian owned corporation established in 1988 to participate in the bid for the project. SCI is the successor company of SCI Engineers and Constructors Inc. SCI is headquartered in Calgary, Alberta, Canada.

Northern Construction has been active in Canadian engineering construction projects for over 40 years. It is a wholly owned subsidiary of Morrison-Knudsen, an engineering construction group, headquartered in Boise, Idaho.

GTMI (Canada) Inc. is a wholly owned subsidiary of GTM Entrepose, a worldwide engineering construction group headquartered in Nanterre, France. GTM Entrepose has expertise and experience in design, construction, project financing and operations.

Ballast Nedum is one of the world's major construction groups. It operates in all sectors of the construction industry, occupying a prominent position in dredging.

Subsequently, Northern Construction dropped out of the consortium.

Questions

1. Quadrant Analysis

    Where does the Northumberland Project fit in the Quadrant framework? Is it a BOT project, or something else?

2. Revenue Based Capital Projections

    Assume the franchise period of 35 years, commencing from June, 1997. Make a reasonable, yet simple, estimate of the annual revenue that you are confident will be received by the franchisee from summing the annual government payment and a conservative estimate of toll revenues. Make an estimate of how much capital can be raised based upon this 35-year stream of annual revenue. Prepare and present a plot of Discount Rate (x-axis) versus Projected Capital (y-axis).

    What other factors would you need to consider to refine this analysis, and to use it as a basic check of the viability of project proposals?

3. Financial Analysis – From SCDI's Perspective

    An investment of this magnitude calls for a reliable financial analysis to establish the viability of the project – first by the government to confirm that prospective bidders will be attracted, and second, by each prospective bidder to confirm that the scope of work can be completed within the financial parameters established by the government in the RFP. Financiers need to see such an analysis prior to making a decision on the amount and terms of loans to finance construction and/or operations.

    This task requires you to carry out such a financial analysis and calculate the Net Present Value (NPV) to SCDI for this investment. For the purposes of this exercise the following assumptions may be made:

    a. Discount rate is 10 %

    b. Annual average CPI growth is 3.0% with 1992 as the base (the 1992 CPI is 128.1 as given in the case material)

c. Annual growth in the amount of traffic is assumed to be 5.0% for commercial vehicles, 3.9% for all other vehicles and 4.0 % for passengers. Further assume that the 1994 traffic figures may be used as a base for these growth calculations.

d. The fixed crossing is expected to generate additional economic activity as also boost tourist inflows. Hence, a one-time increase of 20.0% in 1998, on all forms of traffic, may be taken to result from the fixed crossing.

e. The traffic figures given in the case material are in respect of the traffic leaving PEI.

f. Fares mentioned in the case material may be taken as the initial tolls chargeable in 1997. SCDI can increase this every year at a rate not higher than 75% of the increase in CPI. Tolls are charged only on the traffic leaving PEI, as in the case of the current ferry operations.

g. For simplicity, all passengers may be assumed to be charged a uniform toll of C$ 6.00 and all commercial vehicles are charged C$ 30.00. All other vehicles are charged C$ 19.50. These would be the base year charges of 1997 and liable to escalation annually.

h. Government subsidy is C$ 42 Million per year in 1992 dollars. Hence this would increase annually at the same rate as the CPI.

i. Construction begins in 1993 and is completed in 1997. The capital cost of C$ 840 Million is spent as follows:

Year	1993	1994	1995	1996	1997
Amount	40	200	280	300	20

j. SCDI incurs an expenditure of C$ 1 million during 1997 on operation and maintenance of the bridge. This cost goes up thereafter at the same rate as the CPI.

4. Sensitivity Analysis

A financial analysis provides a base case. Investors would need to be sure of the robustness of their investment proposals and would need to carry out a sensitivity analysis, which would indicate to them the sensitivity of their returns on account of changes in different assumptions or parameters. This task requires you to carry out such a sensitivity analysis and answer some common questions that an investor would like to have.

Do a sensitivity analysis varying the following parameters against the base case:

a. Discount rates of 5%, 8%, 15% and 20 %. At what discount rate is the project no longer viable, i.e. has an NPV = 0.

b. Construction gets delayed by one year and SCDI has to bear the cost of ferry operations. Assume that subsidy on ferry operations grows at the rate of 120% of CPI and has a value of C$ 42 million in 1992. Further the capital cost of construction is spread out to 1998, with spending being 40 Million in 1993 and the balance being equally spread over the next five years.

c. Construction is accelerated such that substantial completion is reached in three years, and the bridge opens for traffic in time for the 1997 tourist travel season.

Year	1993	1994	1995	1996	1997
Amount	40	200	300	300	0

d. Cost of construction goes up by 10% but there is no delay in completion.

e. Cost of construction goes up by 10% and there is also a one-year delay in completion of the project.

f. The one time traffic increase of 20% does not materialize at all.

g. The one time traffic increase is only 10% and not 20%.

What is more critical for the viability of this investment - timely completion or preventing cost escalation?

5. Financial Analysis - Government's perspective

The Government has the option of allowing the ferry operations to continue as before and not take up the bridge. This would mean continuing subsidies at increasing rates. However, if the government decides on the bridge option it can effectively put a cap on its payments in real terms. What, therefore, are the benefits to the government in going ahead with the bridge option. To establish this, the task requires you to carry out a financial analysis from the government's perspective.

Find the NPV of the bridge option for the Government of Canada as compared to the do nothing options. For this purpose the following assumptions may be made:

a. Discount rate is 10%

b. Ferry subsidy is C$42 Million in 1992 and grows at 120% of the CPI increase every year.

c. Other assumptions made in task 1 and 2 may be made for this also. What if the discount rate is 5%.

References

Consolidated Stage III Proposal Call dated May 20, 1992.

21st Annual Statistical Review of Prince Edward Island.

Woods Gordon Management Consultants. Financial Analysis of the Northumberland Strait Crossing Project. May 1987.

Fiander-Good Associates Ltd. Economic Feasibility Assessment for the Northumberland Strait Crossing. July 1987.

Atlantic Provinces Economic Council. Socio-Economic Impact of a fixed crossing to Prince Edward Island. July 1991.

Publicity material of Strait Crossing Development Inc.

"Placing Spans at a Dire Strait." <u>ENR</u> September 16, 1996: 26-30.

"Bridge Over Icy Waters." <u>Boston Globe</u> November 13, 1995: 29-30.

"The Fixed Link." <u>American Scientist</u> Jan-Feb 1997: 10-14.

"Making BOT Work for a Major Bridge Project." <u>PW Financing</u> November 1993: 17-26.

CHAPTER 7   HIGHWAY 407 ETR - TORONTO

Infrastructure Development Systems IDS-97-T-013

Professor John B. Miller prepared this case with the assistance of Research Assistant Om P. Agarwal as the basis for class discussion, and not to illustrate either effective or ineffective handling of infrastructure development related issues. The facts have been intentionally altered for the purpose of considering particular procurement strategy issues in an educational setting. The assistance of Robert W. Gregg of Hughes Transportation Management Systems. Michael W. Mutek, Esq. of Hughes Information Systems, and John T. Kuelbs of Hughes Aircraft Company in the collection of documents relating to the project is gratefully acknowledged.

Introduction

The Highway 407 Central Project is the world's first multi-lane, fully electronic toll highway - no gates, collectors, or speed restrictions - and will extend 69 kilometers across the top of Metro Toronto. The economic center of Ontario is Toronto, a city spread out in an East-West orientation on the North shore of Lake Ontario. A major East-West freeway, Highway 401, and one East-West arterial, the Gardner Expressway, serve the city. Toronto, like the rest of the Province of Ontario, has experienced more than two decades of rapid growth. Over that time, the population of the Greater Toronto Area (GTA) has annually increased by 90,000. Since 1970, the province's population has increased by 20%, more than doubling the number of registered vehicles. Despite an efficient, effective downtown transit system, there seemed to be an insatiable demand for ready movement of people and goods in, around, and through the metropolitan region.

Long term planning for Highway 407 started in the late 1950s. For over 40 years, the metropolitan area plans have called for another East-West highway, to the north of Highway 401, on land reserved for just such a purpose. Throughout these four decades, Highway 401 has become the second most congested highway in North America. The proposed Highway

407 has become an important element in the general need to ease traffic congestion on Highway 401 through an alternate route. Except for a few bridge structures carrying North-South road crossings along the route, successive Ontario governments were unable to dedicate enough funding to deliver Highway 407, despite studies that have put the cost, in lost time and productivity, to the GTA business community of traffic delays at $C2 Billion annually. Chronic roadway congestion on Highway 407 is also an important contributor to air pollution in the region.

Paul Decker, Hughes

Paul Decker, the Program Manager on the Highway 407 ETR project for Hughes Aircraft of Canada Limited ("HACL"), was excited about the upcoming completion of the project. Hughes and its team of technology integrators would introduce a step-change in management of tollways throughout the world. Paul grew up in Toronto, and still found it to be the most attractive urban centers in North America. Paul attended local public schools before college, where he majored in civil and environmental engineering. He had ten years of experience with BOT consortiums in Hong Kong, which was very helpful during the Highway 407 competition. Paul was familiar with the extraordinary effort required to integrate design, construction, technology, maintenance, operations, and finance within a proposal team whose goal was to win a competition for exactly this kind of project.

Throughout his Hong Kong experience with Design Build Operate and Build Operate Transfer competitions, innovation was ruthlessly sought in ways fundamentally different than in Design Bid Build. Competition occurred later in the procurement process and covered a significantly enlarged scope. (See Figures 7-1 and 7-2.) Tenderers were required to compete for combined functions of design, finance, construction, maintenance, and operations, a competition that expressly values innovation in each of these areas and in the integration of one or more of these areas. DBO/BOT delivery methods put tenderers in the position of system integrators whose role was to present alternative combinations of quality, time, and life-cycle cost for a project that meet the Owner's functional needs. Tenderers perform this "system integrator" function <u>before</u> the Owner makes a firm commitment to any one package. Paul found the process to challenge every member of the proposal team to help produce this "synergistic" combination of quality, time, and life-cycle cost. Paul had seen unusually creative solutions emerge from this process in Hong Kong's Tate's Cairn Tunnel, Eastern Harbor Crossing, and Western Harbor Crossing. What drove each competing consortium was the knowledge that each team would inevitably package basic project elements differently, with resulting tradeoffs in quality, cost, and time. Paul's experience was that this competitive pressure produced new applications of technology, new combinations of design and construction methods, and innovative ways to produce higher quality, to produce greater life cycle cost savings, and to produce additional savings in time.

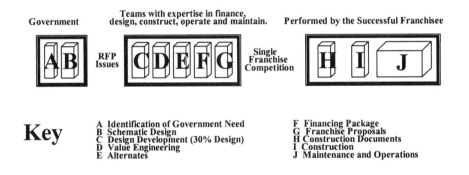

Figure 7-1   Typical DBO and BOT Schematic

Multiple evaluation factors were often the natural result of these processes -- such as time to revenue, time to completion, interest rate, life cycle cost, initial capital cost, and long term maintenance and operations costs. Paul found that such factors drove competitors toward a kind of "balanced excellence" he found refreshing. The Hong Kong government benefited from a much richer choice among fully functional projects, while the competing consortia learned a great deal about innovative approaches to satisfying integrated customer needs. Establishing evaluation factors prior to the issuance of an RFP was more complex than Design-Bid-Build, but the trade-offs were clear.

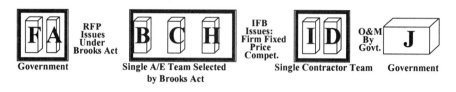

Key

A Identification of Government Need   F Financing Package
B Schematic Design   G Franchise Proposals
C Design Development (30% Design)   H Construction Documents
D Value Engineering   I Construction
E Alternates   J Maintenance and Operations

Figure 7-2   *Design Bid Build Schematic (U.S.)*

Paul was looking forward to the May 15th declaration of "victory" on this $1 Billion project. The project was expected to open for public operation in just three weeks, which happened also to be Paul's 42nd birthday. He was ready to move on to other AVI/ETC projects about to be advertised around the world. HACL had managed an extraordinary team of technology suppliers on Highway 407, including Bell Canada, Bell Sygma Telecom Systems, Mark IV Industries Ltd., Delco Electronics, Nortel, Fore Systems, CTI Datacom, Hughes Missile Company, and Advance Toll Management Corporation ("BBMH"). Each had worked very hard since November 1994 to produce the AVI/ETC systems for Highway 407 ETR. Unless and until traffic flowed and tolls were collected, there would be no receipts for Hughes or other members of the team, since all payments were tied to collection of tolls. Paul was convinced that the technology they had developed would offer the kind of competitive advantage that could sustain Hughes' Transportation Management Systems Division for a decade.

A New Assignment

At 6:30 PM, Paul began packing up for the day -- thinking about Singapore, Kuala Lampur, Israel, and Australia. All were likely to conduct AVI-ETC competitions in the next eighteen months. But first things first.

The Hale Bopp comet was on the tonight's agenda for Paul, his wife, and their three kids. Before he got out the door, the phone rang. It was Mike Roberts, head of Hughes' Transportation Management Systems in Fullerton, California, the ETC prime contractor on Highway 407, and sister company to HACL. Roberts' group had provided all the Toll Transaction Processors (TTP's) and the Roadside Toll Collection System (RTC's) for Highway 407, the guts of Hughes' technology base for ETC.

"Glad I caught you, Paul," Mike began. "I need your help to complete an assignment from Hughes corporate. I've been asked to summarize what we learned from participating in the Highway 407 ETR project. The bottom line assignment is an evaluation of the civilian government market for Hughes' ETC technology. Assuming that we have a leading edge toll road technology, what did we learn about how best to deploy that technology in future procurements throughout the world?

"I'd be glad to help, but I need to head out early tonight to go star -- I mean -- comet gazing," Paul replied. "I'll try to get some thoughts back to you next week."

Mike was relieved. "I'm worried about these next several solicitations for AVI/ETC projects. Several public owners say they are using the Toronto procurement as a model. We'll be presenting our views to a strategy meeting in Long Beach at the end of the month, which will include our teaming partners from Toronto."

Paul and his wife, the Toronto business correspondent for the New York Times, with an MBA from Michigan, explained stargazing, Hale-Bopp, and comets to their kids that night. On the way home, Paul described his conversation with Mike Roberts. What had seemed, at first, like a relatively straightforward confirmation of the superiority of Hughes' technology, now seemed a bit like stargazing. Being asked to evaluate the Toronto procurement process for top corporate management added a little excitement -- both positive and negative -- to the task. He'd have to review his files on the project, and reorient himself to thinking more broadly about Highway 407.

The History of Private Sector Involvement in Highway 407 *ETR*

In 1993, the then Liberal (the name of a political party in Canada, not an ideological description) government of Ontario decided to accelerate the construction of Highway 407 by inviting the private sector to design, build, operate and finance it. Figure 7-3 shows the layout of the new Highway 407 ETR. Government's purpose was to reduce the burden on the taxpayer and to expedite the construction of the facility. The Government decided to develop the project as a toll road, and created a new Crown agency, the Ontario Transportation Capital Corporation (OTCC) to complete large transportation projects quickly and cost efficiently by working in "partnership" with the private sector.

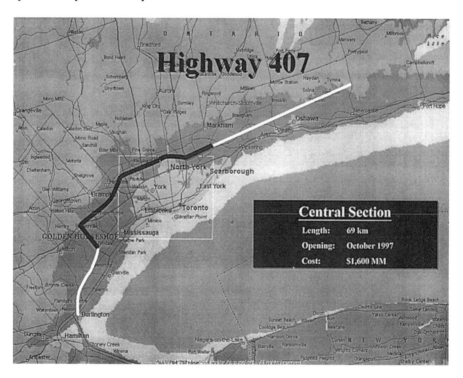

Figure 7-3   *Map of the Highway 407 ETR*

The procurement plan for the Highway 407 Project originally focused upon a BOT process. Billed as a job creation effort, the Province issued a Request for Qualifications in May 1993 to two pre-selected consortia for the BOT development of Highway 407 as an electronic toll road. The government contracted with each group in the amount of $1,500,000 each for "value engineering assessment" reports which identified cost effective design and construction alternatives to existing Ministry of Transportation, Ontario ("MTO") standards. Based upon these two reports, the MTO issued broad, baseline design/construction standards for the 407 project. The standards did not specify the number of lanes, type of pavement, type and extent of lighting, or means for electronic toll collection. The MTO issued its final RFP in September 1993, for detailed proposals and business plans to finance, design, build, maintain, and operate Highway 407 as a toll highway for 30 years. Each team was to make segregable proposals for the road and the ETC facilities, with full descriptions of financing plans. Exhibit 7-1 contains relevant portions of the RFP. Exhibit 7-2 contains a modified version of the teaming agreement used by one of the ETC consortia as a basis for pursuing the work described in the final RFP.

Exhibit 7-1   *Excerpts from the RFP*
Exhibit 7-2   *ETC Teaming Agreement*

The project generally involved conventional design and construction issues associated with the construction of a large interstate highway. Integrating such a conventional design with advanced AVI/ETC systems for managing traffic and tolls proved to require more innovation and creativity than any of the project's planners had foreseen. In developing their respective proposals, the two competing consortia quickly discovered that there was a direct inter-relationship between the AVI/ETC system used and highway exit/entrance ramp configurations. For example, one of the proposed AVI/ETC systems required no toll plazas whatsoever. As a result, the configuration of access and egress ramps could be quite similar to that for a free interstate highway common throughout North America. The other AVI/ETC system required both transponders and toll plazas for the collection of tolls. This electronics package led to different treatment of access and egress ramps that could accommodate lane changes at interchanges, along with accelerating and decelerating traffic at toll booths.

In December 1993, the two consortia submitted their tenders, which produced dramatically different results.   Table 7-1 summarizes the proposals.

Item	Group 1 Tender	Group 2 Tender
Road Lanes	4-6 initial lanes, expandable to 6-10 lanes	4 initial lanes, expandable to 8 lanes
Lighting	58 Km fully illuminated, 11 km partially illuminated	Certain interchanges illuminated
Pavement	Concrete pavement (30 year estimated design life)	Asphalt 7-10 year design life, Rehabilitation begins in 2003
ETC	Mixed toll collection system, with automatic vehicle ID and manual toll booths (MFS Technologies)	Fully electronic tolling system with video-tracking (Hughes/Bell)
Schedule	Complete by 1999	Complete by 1997
Project Finance	Primarily Debt	Primarily Debt
	Nominal Equity Contribution	Nominal Equity Contribution
	Substantial Govt. Subsidy	Substantial Govt. Subsidy

Table 7-1   Comparison of the Highway 407 Tenders
Source: Annual Report of the Provincial Auditor

After evaluating the competing proposals, the Government made the surprising decision to un-bundle the BOT procurement. The government announced in April 1994 that it had selected Group 1's proposal, except for the tolling component. Six weeks later, the government announced that the ETC team from Group 2 had been selected to supply the toll system. In sum, the government selected the road proposal from Group 1 and the ETC proposal from Group 2. Financing was removed as one of the evaluation factors, and the road portion of the project was converted to directly financed Design-Build. The ETC portion of the procurement was converted to Design-Build-Operate.   The stated justifications for changing procurement approach were, first, that neither team put a substantial amount of its own equity at risk, relying almost exclusively on debt.   The Government reasoned that it could finance the same debt at a substantially lower "cost of money". The second rationale related to the quality of basic project elements. Given the decision to finance the project with government debt, the Government decided that taking the higher quality road elements from Group 1 and the higher technology ETC elements from Group 2 offered the best value to the Province. The Government intervened to break both the teaming arrangements that had been signed within each consortium to allow the toll system team from the losing consortium to provide its technology.   Each consortium had its tolling suppliers tied up with an exclusive teaming arrangement.

Breakthrough Technology in Toll Collection: AVI/ETC on Highway 407 ETR

The Ontario government sought advanced technology for toll collection.   "The toll collection systems shall be capable of accurately detecting and identifying vehicles under all operating conditions with an accuracy of 99.995%."

Each user of the road passes under two gantries fitted with a vehicle detector, license plate cameras, and read/write UHF antennas. The vehicle detector identifies the class of each passing vehicle (e.g. automobile, truck) and the antennas determine whether each passing vehicle has a transponder. For those with transponders, the antennas read the appropriate account number, the corresponding vehicle classification, and write the point of entry, the time and date of each vehicle, via fiber optic cable, onto a toll processing center computer for assessment against the user's account. For those without transponders, the license plate camera photographs the rear of each vehicle as it enters or exist Highway 407 ETR and sends a digitized image of the license plate to the toll processing center, for assessment and billing to the owner of the vehicle.

The Highway 407 ETC systems utilize the Slotted Aloha, Time Division Multiple Access (TDMA) protocol, originally developed at the University of Hawaii for military communications. The TDMA protocol allows concurrent communication at more than 500,000 bps with up to 272 vehicles on an open road. The protocol can identify and assess toll charges for vehicles traveling well in excess of 100 mph by locating vehicle transponders spatially to within fractions of a meter. Multiple motorcycles traveling in a single lane can be monitored individually, even when they are nearly touching. This is accomplished using Hughes' patented Angle of Arrival (AOA) technique, which uses a phase interferometer to precisely measure the position of specific transponders at a place and time for toll collection purposes. A variety of payment methods make collections a snap.

The Second Teaming Agreement. Shortly after Group 2 technology team (known as "BBMH") was selected to be the AVI/ETC supplier for the project, Paul Decker and representatives from a number of other companies signed a second Teaming Agreement. Pursuant to this teaming agreement, Hughes and its team were to produce and deploy the contemplated AVI/ETC system.

The Provincial Auditor's Report

In its 1996 Annual Report, the Office of the Provincial Auditor released an interim audit report for the project, with the following overall findings:

Overall Audit Observations

The Highway 407 Central project was structured to be a new delivery model for the construction of Ontario highways. According to this model, the highway was to be designed and built for a fixed price by a private sector contractor with construction costs to be repaid through tolls paid by users of the highway. This arrangement, especially the collection-of-tolls aspects, allows for larger projects to be undertaken at a faster pace.

The timetable for Highway 407 Central was an ambitious one when compared to traditional provincial highway construction timetables.

Much importance was placed by the government on the urgent need for a new highway to relieve traffic congestion in the Metropolitan Toronto area, to take advantage of the low business cycle with potentially lower construction costs, and to stimulate the economy by creating employment in the road construction sector.

We observed that, although cited as a public-private partnership, the government's financial, ownership, and operation risks are so significant compared to the contracted risks assumed by the private sector that, in our opinion, a public-private partnership was not established.

We found that the selection of the winning submissions for the design and construction of the road and tolling components of the project followed a pre-determined process and the proposals were evaluated by experienced evaluation teams using pre-determined evaluation criteria. An external firm of management consultants was hired to oversee the selection process.

We concluded that there was due regard for economy and efficiency in the planning, development and implementation of the Highway 407 Central project to the extent indicated by the points below.

- There was a demonstrated need for Highway 407.

- Value engineering was used in designing the highway.

- Economic processes were in place for project monitoring as to the quality and progress of the winning consortium's work.

- A number of the request for proposal objectives were achieved or are in process.

However, we found that there are areas the Ministry of Transportation needs to consider for improvement when undertaking projects of a similar nature in the future. Specifically, these areas are:

The minimum number of bidders and design and construction alternatives needed to provide an adequate basis for decision-making. In the case of Highway 407 Central, only two consortia submitted two design and construction proposals;

The Highway 407 Central request for proposal provided very general design criteria. For example, the number of lanes, type of pavement and type of illumination were not specified. The Ministry needs to weigh the benefits gained from providing the private sector with the flexibility to be innovative against the cost of having bids that may not be price-comparable because they are so different;

The clarity of the request for proposal in conveying to the bidders the Ministry's intentions and expectations regarding the sharing of risks and rewards;

Whether components of a project that have become separated or "unbundled" from the original request for proposal need to be tendered separately. The Highway 407 Central request for proposal challenged the two consortia to submit integrated proposals taking on all risks associated with the highway including financing and associated risks, construction, and operations and maintenance over a 30-year period. Ultimately, the project was not fully integrated as the province assumed 100% of the financing and operating risks. The removal of private financing from the Highway 407 Central project meant that the Ministry would be responsible for financing the project and would assume operating and ownership risks from the outset rather than after 30 years as originally envisioned.

The tolling system was unbundled and awarded separately to the tolling consortium of the losing proponent. As well, the operations and maintenance contracts were to be awarded to the winning consortium without separate tenders. Apparently, it would have been feasible to separately tender both the highway maintenance and tolling system contracts.

Paul's Report to Hughes Senior Executives

Over the weekend, Paul thought how best to describe the lessons learned from Toronto, and in particular, how to apply them to five international projects that were expected by the summer of 1997.

Questions

1. What are the five most important procurement related events to include in his summary to the executives at Hughes, in chronological order. How did each event affect the parties (MTO, Design/Builder, the BBMH joint-venture), the process, and the result?

2. Where did this procurement fit in the Four Quadrants, initially, and at the end?

3. Was the government's decision to un-bundle the procurement justified by the low amount of equity offered by each consortia? Point to the portions of the RFP that signaled to the competitors what the government's intentions were with respect to the impact of low equity on its evaluation.

   Hughes' executives have asked how this apparent miscommunication between the proposers and the MTO arose.   Based upon the procurement process and the text of the RFP, present the logic by which the original consortia each independently decided to propose very little equity into the road side of the project, and to finance the original project (BOT) primarily through debt? Did the parties respond to the RFP as the government should have expected? Why or why not?

4. Separately for each, what should Paul, Hughes, and the MTO have learned from the procurement process for Highway 407?

   Paul is well familiar with the Northumberland Strait Crossing Project. Compare the revenue streams associated with the original Highway 407 project to those associated with the original (and only) RFP for the Northumberland Strait Crossing.

5. Are the Provincial Auditor's Overall Findings properly focused upon the key issues associated with this procurement? What suggestions, if any, would you make to expand, contract, or redirect their review?

6. Evaluate the sections of the RFP, Exhibit 7-1, which are found on the companion CD.

How did these sections influence the ways in which the proposers and the MTO conducted themselves in this procurement?

    a.   The evaluation criteria in Section 7.0

    b.   The Principles set forth in Section 5.2

    c.   Financing Plan requirements in Section 5.4

    d.   Section 5.9, Procurement of Design and Construction Services

    e.   Section 3.14, Forecasted Traffic Volumes, and

    f.   Sections 9.1.6 and 9.7,

    g.   Section 9.10 (Ownership and Copyright) and Schedule 7, and

    h.   Any Other parts of the RFP you believe to be important in shaping private sector interest in the project.

References

Ng, Joseph S., Stephen R. Sherman, and David W. Lester. "ATM Streamlines Toronto's 407." ITS International Issue 4, March 1996.

McDaniel, Thomas, Dennis Galange. "Highway 407 Sets Standard for ETC." Traffic Technology International 96, Issue 4, March 1996.

Hazan, Ralph, Alex Castro, Jr., and A. Martin Gray. "The Compatibility Factor." ITS International Issue 5, June 1996.

Stephen H. Daniels. "Smart Highway Set for IQ Test." Engineering News Record January 27, 1997.

"Canada Project Speeding Along." Engineering News Record August 26, 1996.

The Provincial Auditor of Ontario, Canada. Annual Report of the Provincial Auditor 1996: 234 et seq.

CHAPTER 8   HUDSON-BERGEN LIGHT RAIL TRANSIT SYSTEM

Infrastructure Development Systems IDS-00-T-019

Research Assistant Kai Wang prepared this case under the supervision of Professor John B. Miller as the basis for class discussion, and not to illustrate either effective or ineffective handling of infrastructure development related issues. Data presented in the case may have been altered to simplify, focus, and to preserve individual confidentiality. The assistance of John Johnston, Thomas Wood, the staff at NJ TRANSIT, and the staff at the 21st Century Rail Corporation are gratefully acknowledged.

Introduction

Hudson and Bergen counties are located along the Hudson River at the north side of New Jersey. From this vantage point, the skyline of Downtown Manhattan seems just a stone's throw away. In 1988, the New Jersey Transit Corporation (NJ TRANSIT) conducted a study to assess transit needs in the area. This study, funded through a federal government grant, considered alternatives and their impact on the environment. The report concluded that a new light rail system connecting Hudson and Bergen counties in a north-south direction along the waterfront would best promote the development of the area. Existing service in the area -- bus, ferry, rail, and subway –generally served east-west commuter movements, and the new line would facilitate movements along the waterfront.

Exhibit 8-1 shows the proposed route of the new Hudson-Bergen Light Rail Transit system (the "HBLRT"). Exhibit 8-2 shows an artist's rendering of how the new system will appear, when finished.

Exhibit 8-1   *The Proposed Route of the Hudson Bergen Light Rail Transit System*
Exhibit 8-2   *Artist's Rendering of the HBLRT in Operation*
Exhibits 8-1 and 8-2 are Courtesy of 21st Century Rail Corp.

The New Jersey Waterfront is one of the most densely populated areas in the U.S., and is expected to experience considerable residential, retail, and commercial redevelopment. Numerous development sites are popping up in areas such as Bayonne, Jersey City, Hoboken, and Weehauken. Housing units along the shore have grown to more than 8,500 units with rental values ranging from $950 for a studio to $3,000 for a three-bedroom duplex per month. Notable retail ventures like the Newport Center Mall, with nearly 1 million square feet of retail space, are being developed in the area. Several Fortune 500 Companies such as Lehman Brothers, Paine Webber, and Merrill Lynch, have established substantial offices along the waterfront. As the skyline along the New Jersey side of the Hudson River continues to expand, the HBLRT is expected to play an important role in complementing the accessibility needs of the area.

The HBLRT system will accommodate 100,000 passengers per day along the waterfront by 2010. Initial ridership is projected at 25,000 passengers per day, in 2001. The new rail system is expected to reduce commuting time by up to half an hour per day.

The NJ Transit Corporation

The New Jersey Transit Corporation ("Corporation") was formed in 1979 by the New Jersey State Legislature to manage the development and maintenance of the state's public transportation systems. This included the majority of the bus routes in the state, the Newark City Subway System, and twelve commuter rail lines. NJ TRANSIT is the nation's third largest bus and rail service provider, serving nearly 300,000 daily passengers over 5,400 square miles of land.

The Corporation's Operating Funds

NJ TRANSIT's operating funds are generated from fares charged to riders, state and federal subsidies, and other revenues generated within the system. In 1995, the NJ TRANSIT collected $535.8 million of its $814 million operating budget (i.e. 66% of operating expenses) from fares and system-generated revenues. The balance of the Corporation's budget (34%

of total expenses) represent subsidies received as federal and state aid. Despite a freeze in State subsidies and a decrease in Federal operation subsidies in the early 1990's, NJ TRANSIT had managed to improve on-time performance of its network without increasing the prices of the fare for the four years preceding 1995. Average on-time performance was over 94%, and passengers appear to be aware of these improvements. Unlike other transit systems, the number of NJ TRANSIT riders was increasing throughout the period that HBLRT was being planned. In 1994, the Corporation received the Transit System of the Year Award from the American Public Transportation Association (APTA).

The Corporation's Capital Program

None of the funding for the Corporation's capital program comes from fares or system-generated revenues, which are sufficient only to cover two-thirds of operating expenses. Yet, like most major transportation authorities in the United States, the Corporation has a significant ongoing capital program that includes capital repair and maintenance of existing assets and the staged implementation of four major new capital projects valued at over $2 billion. Balancing the need for ongoing capital repair of existing facilities with the need for new facilities is a constant, and difficult problem for the Corporation to manage.

NJ TRANSIT's Fiscal Year 1995 capital program budget was $490.65 million. Of that amount, 59% ($289.48 million) was provided by the federal government and 41% ($201.17 million) by the State. The capital program is approved on a 5-year "look-ahead" basis, but reviewed and revised annually, as appropriate. The program includes design and construction of bus maintenance facilities, rail station rehabilitation, and the procurement of bus and rail rolling stock.

Federal grants come from three main sources. In FY1995, approximately $95 million of the appropriation came from the Federal Transit Administration (FTA) Section 9 program. Approximately $65 million came from the FTA's Section 3 Fixed Guideway Modernization Program. The remainder – approximately $330 million – was appropriated by Congress to the Corporation through the "Urban Core" program, a collection of projects intended to integrate the mass transit network in northern New Jersey. Currently listed in the Urban Core program are the following projects:

- Kearny Connection – this route connects NJ TRANSIT's Morris & Essex Line to Amtrak's Northeast Corridor. The connection will run though Hoboken and mid-town Manhattan. (Budget: $60 million, Scheduled completion, 1996).

- Secaucus Transfer – this new rail station along Amtrak's Northeast Corridor will allow passenger to transfer among the Main, Coast, and Northeast Corridor lines. (Budget: $400 million, Scheduled completion, 1998).

- Hudson-Bergen Light Rail Transit system – the subject of this case study. (Budget: $1.2 billion, Scheduled completion of first phase, 2000).

- Newark-Elizabeth Rail Link – a new light rail line from downtown Elizabeth through Newark Airport to Belleville/Nutley. (Budget: $600 million, Scheduled completion, 2004).

The cost of these four projects alone – over $2.2 billion – far exceeds the amount of funds expected from the federal government over the next ten years. The amount of federal grant funds is uncertain from year to year, because the existence of a full-funding agreement between the Corporation and FTA is unconnected to actual appropriations in any given year.

State grants in support of NJ TRANSIT's capital program are provided through the New Jersey Transportation Trust Fund (TTF). This fund is primarily underwritten by gasoline taxes collected from automobile users and from contract payments made by state toll road authorities. Each year, the New Jersey Legislature appropriates funds from the TTF to specific projects from a list submitted by the NJ Commissioner of Transportation. It is anticipated that the TTF program will provide approximately $258 million per year into NJ TRANSIT projects. For FY 1996, $28 million was appropriated to the HBLRT from this fund.

The Corporation's Capital Program is also supported by appropriations of federal highway funds to the New Jersey Department of Transportation (NJDOT). The amount of funds received by NJ TRANSIT is dependent, from year to year, on the amount received from federal highway funds, and upon the allocation of such funds by NJDOT to a variety of uses, including NJ TRANSIT. Unpredictably, miscellaneous funds are available to fund NJ TRANSIT projects, often in cooperation with other agencies, such as the Port Authority of New York and New Jersey.

The Original Procurement Plan

NJ TRANSIT originally planned to deliver the project using the Design-Bid-Build method, in which a professional engineering team would completely design the project and deliver completed plans and specifications to NJTRANSIT. The Corporation would the use these plans and specifications to conduct a competition for a construction contractor to furnish and install all the equipment and materials specified in the design for a firm fixed price.

The drawbacks to this approach were several: first, the Corporation believed it was likely that the project would have to be broken up into several construction contracts. The award of multiple construction contracts would create a substantial coordination problem both during the design phase, and during the construction phase. It seemed unlikely that multiple contracts could be advertised, awarded, commenced, and completed in accordance with a single schedule established by the Corporation. If delays, changes, or differing site conditions were encountered, they would likely be dispersed among the contracts, making it very difficult for the Corporation to coordinate the completion of multiple contracts. Substantial delay in even one segment of the new transit line could have a critical domino effect on the ability of major equipment suppliers – particularly the transit car supplier and the electronic control system supplier – to install and test their equipment over the entire system.

Second, multiple design-bid-build contracts were likely to take much longer to complete. The Corporation believed it likely that costs would escalate over time, which was likely to create a substantial problem for the Corporation over the many years of design and construction.

Third, the Corporation wanted to achieve the earliest possible opening for the system, since north-south transit needs were significant, and user benefits would be high, once the project was completed.

The passage, in 1991, of the Intermodal Surface Transportation Act (ISTEA) encouraged public owners to experiment with "innovative" project delivery methods. In 1994, NY TRANSIT decided to change the delivery method for the project from Design-Bid-Build to either Design-Build-Finance-Operate or Design-Build-Operate. Booz-Allen & Hamilton was retained by the Corporation to provide overall program management for the project.

Figure 8-1 shows: (a) how the HBLRT was original intended to fit into the Four Quadrants, and (b) the two project delivery methods subsequently considered by the Authority.

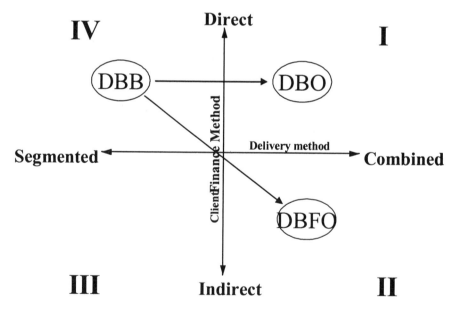

Figure 8-1   HBLRT Procurement Strategies in the Four Quadrants

Technical Definition of the HBLRT Project

NY TRANSIT relied on the many previous studies to establish the technical definition of the Project. HBLRT would be a double track, light rail system extending for 20.5 miles from Bayonne to Ridgefield. The HBLRT system would include thirty-two (32) stations and five (5) park-and-ride lots. Construction of the project was to proceed in three segments.

- In the first segment – the "Initial Operating Segment (IOS)" – 9.5 miles of rail would be laid from Bayonne, through Jersey City, and through Hoboken. (See, Exhibit 8-1, above.) Upon the opening of the first segment, twenty-nine (29) Light Rail Vehicles (LRT were to be operational at the time this segment. The first part of the IOS, from Bayonne to Exchange Plaza in Jersey City, is expected to be operational in March 2000. The second part of ISO is expected to open in December 2000, extending the system from Exchange Plaza north to Newport Station. The last part of the IOS is expected to open in mid-2001, linking Newport Station to Hoboken Terminal.

- In the second segment – the Subsequent Operating Segment (SOS) – the route will be extended in both directions: north from Hoboken Terminal to North Bergen and further south within Bayonne.

- In the third segment -- the Final Operating Segment (FOS) – the project will extend the system yet further north to the Vince Lombardi Park-and-Ride in Ridgefield.

Commercial Definition of the HBLRT Project

When NJ TRANSIT switched the project delivery method from Design-Bid-Build (DBB) to Design-Build-Operate (DBO) or Design-Build-Finance-Operate (DBFO), the commercial terms under which the private sector would furnish and install the materials, equipment, and services associated with the HBLRT changed significantly. With the use of the integrated delivery methods (DBO or DBFO), NJ TRANSIT required the contractor to provide the design, the equipment, the materials, and both operation and maintenance services for the system for an initial term of 15 years. NY TRANSIT also required the contractor to furnish and install sixteen (16) low-floor Light Rail Transit (LRT) cars for the Newark City Subway, and to design and construct a new Newark City Subway Vehicle Base Facility (NCSVBF), linked to the Newark City Subway system to maintain the LRT cars.

Covering the Cost of Initial Delivery

In an attempt to provide a level subsidy to the HBLRT project over the construction period $400 million was allocated to the project for FY1996 through FY2000 in five equal installments of $80 million per year, including NJ TRANSIT's expenses. See Table 8-1. Table 8-2 shows how much of this $80 million is expected to be actually available for to pay for design and construction of the system during each of the five years of construction related activity.

	1994 Actual	1995 Actual	1996 Estimate	1997 Estimate	1998 Estimate	1999 Estimate	2000 Estimate
Total Program	486.0	490.65	622.9	557.4	570.8	584.8	759.1
HBLRT	12.4	26.2	80.0	80.0	80.0	80.0	80.0
% of Total	3%	5%	13%	14%	14%	14%	11%
Total Program Composition							
Federal	59%	59%	54%	NA	NA	NA	NA
State	40%	41%	41%	NA	NA	NA	NA
Other	1%	0%	5%	NA	NA	NA	NA

Table 8-1   NJ Transit Capital Budget Estimates

Fiscal Year	NJ TRANSIT Capital Budget	NJ TRANSIT Costs	Available For DBOM Contractor
1996	$80.00	$16.97	$63.03
1997	$80.00	$38.30	$41.70
1998	$80.00	$24.04	$55.96
1999	$80.00	$10.18	$69.82
2000	$80.00	$9.44	$70.56
Total	**$400.00**	**$98.93**	**$301.07**

(Millions of Inflated Dollars)

Table 8-2   Capital Allocations During Construction to the HBLRT

The contractor was expected to provide whatever short-term financing is needed to bridge the gap between available funds from (state and federal) grants and actual costs incurred during the design and construction period. Figure 8-2 shows schematically, but not in detail, how the State's five annual payments of $80 million dollars are expected to produce surpluses in the first and last year of construction, but shortfalls in years 2, 3, and 4. The Commercial terms of the RFP require the contractor to use the surpluses (and interest earned) thereon to finance the shortfalls.

The State assumed in the RFP that these payments were sufficient to cover both the actual cost of design and construction and any financing costs associated with adjusting the State's level payment commitment to actual cash flow requirements of the contractor.

Covering Fifteen Years of Operation

Operating revenues from the HBLRT will be collected by NJ TRANSIT, including fares, parking fees, and fees generated from advertising agreements, concessions, and other activities. This revenue will be used to fund payments to the contractor for operating and maintaining the system over the first fifteen years of operation. NJ TRANSIT made the commitment, in the Request for Proposals, to provide whatever additional funds are necessary to make the operations contract payment to the contractor, if HBLRT operating revenues are insufficient.

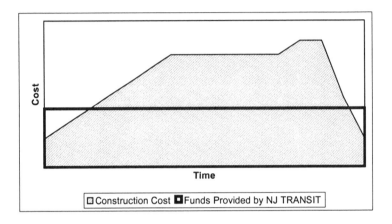

Figure 8-2   *Short-Term Financing Component of the HBLRT Project (Schematic*

Although this commitment was in writing, supplemental funds would still have to be appropriated by vote of the New Jersey Legislature. There is a long experience in the State Legislature of regular appropriations to NJ Transit for operations of approximately $250 million per year. There is a similar long experience with the federal government of regular appropriations of approximately $30 million annually to NJ TRANSIT for operations. If these appropriations were unavailable, earmarked elsewhere, or otherwise insufficient, NJ TRANSIT would then allocate additional funds from its annual TTF funds to cover the remaining operating deficits.

The Procurement

NJ TRANSIT's procurement process consisted of a Request for Qualifications (RFQ), followed by a period of "pre-solicitation" document review leading to the submission of Final Proposals.

The RFQ was issued on March 2, 1995. Responses that submitted qualifications were received from 5 consortia:

- Garden State Transit Group
- Waterfront Transit Consortium
- NJ Translink
- RIK Consortium, and
- Gateway Transit Associates.

Table 8-3 lists the members of each joint venture (consortium) that responded to the RFQ.

Consortium	Members
Garden State Transit Group	Flour Daniel, Inc.
	Siemens Transportation System
	Daniel, Mann, Johnson, & Mendenhall, Inc.
	Hill International, Inc.
	Herzog Transit Services, Inc.
Waterfront Transit Consortium	Bombardier Transit Corporation
	Perini Corporation/Slattery Associates, J.V.
	STV Construction Services
	Alternate Concepts, Inc.
NJ Translink	ABB Traction, Inc.
	Bechtel Corporation
	New York Waterways
RIK Consortium	Raytheon Infrastructure Services, Inc.
	Kinkisharyo (USA) Inc.
	Itochu Rail Car Incorporated
Gateway Transit Associates	AEG Transportation Systems
	Yonkers Contracting/Granite Construction
	ICF Kaiser
	DeLeuw Cather
	Academy Bus Tours

Table 8-3   Membership in the Joint Ventures Responding to the HBLRT RFQ

During the pre-solicitation document review process, ABB Traction, Inc. and AEG Transportation Systems merged. As a result, Gateway Transit Associates withdrew from the competition in favor of the team of which ABB was a part.

Prior to the date for submission of final proposals, Waterfront Transit Consortium withdrew from the competition, claiming that the fifteen year operating and maintenance period was excessive and made the project unattractive.

Three members of the Waterfront Team then joined the Raytheon led team as subcontractors.

Final proposals were submitted by:

- 21st Century Rail Corporation (Originally RIK Consortium),
- Garden State Transit Group, and
- NJ Translink.

Table 8-4 lists the members of each joint venture (consortium) that submitted final proposals.

Consortium	Members
21st Century Rail Corporation (RIK Consortium)	Raytheon Infrastructure Services, Inc. (RISI)
	Kinkisharyo (USA) Inc.
	Itochu Rail Car Incorporated
Garden State Transit Group	Flour Daniel, Inc.
	Siemens Transportation System
	Daniel, Mann, Johnson, & Mendenhall, Inc.
	Herzog Transit Services, Inc.
NJ Translink	Adtranz
	Bechtel Corporation
	New York Waterways

Table 8-4   Membership in the Final Joint Ventures

Proposal Evaluation and Selection of the Raytheon Joint Venture

NJ Translink's final proposal was rejected by NJ TRANSIT as non-responsive. The Corporation took the position that Translink's failure to provide a proposal for more than six years of operations and maintenance of the system required that the proposal be rejected without further evaluation.

The two remaining proposals were evaluated on price and technical approach. The RFP had assigned 60% of the evaluation points to price and 40% to technical approach. Within the 40% assigned to technical approach, the Corporation applied five sub-factors to each proposal, but the description of each sub-factor and the weights assigned to each sub-factor were not disclosed to the competitors prior to the date proposals were due. The proposals are described as "bids" in the RFP. The amount of technical evaluation points assigned to each sub-factor is shown in Exhibit 8-3. The entire RFP section on "Evaluation Criteria" is also reproduced as Exhibit 8-4 on the companion CD.

| Exhibit 8-3 | *Subfactors Applied by NJ TRANSIT in the HBLRT Competition* |
| Exhibit 8-4 | *Evaluation Criteria (pp II-90 to II-95)* |

The results of NJT Technical and Price Evaluation were reprinted to the NJT Board via a document entitled "Board Item Fact Sheet" which is reproduced as Exhibit 8-5 on the companion CD.

| Exhibit 8-5 | *Board Item Fact Sheet* |

Price Evaluation

Garden State Transit Group's price for design, construction, and 15 years operation was $1.8 billion. The price submitted by the joint venture led by Raytheon – 21st Century Rail Corporation – was remarkably lower at $1.26 billion. Seventy percent (70%) of the 21st Century Rail Corporation's stock was owned by Raytheon. The remaining interest was held by the joint venture of Itochu Rail Car and Kinkisharyo USA. See Exhibits 8-6 and 8-7 provide more on the Structure and Background of 21st Century Rail Corporation.

Exhibit 8-6   *Structure and Background of 21st Century Rail Corporation*
Exhibit 8-7   *"Bids Hundreds of Millions Apart" Reprinted from Engineering News-Record, copyright The McGraw-Hill Companies, Inc., July 1, 1996, All Rights Reserved.*

The price submitted by 21st Century Rail Corporation included the following:

- 40 month Design/Build of the transit system: $724 million, (Including design, construction, testing of the system, and manufacture and commission of the light rail vehicles),

- 15 years of operations and maintenance: $435 million, and

- furnish and install the Newark City subway extension and vehicles: $100 million.

John Johnston, President and CEO of 21st Century Rail Corporation, explained that they were able to achieve significant savings in their proposal through value engineering and very strong financial planning. The team proposed to pay off all the project's financing in only eight (8) years, which produced a significant savings in financing costs. NJT's expected price was not disclosed.

A Financing Change

In the Project Prospectus, included in the RFP by NJT, the Authority reserved the right to: accept the offer as proposed; seek more favorable terms based upon changes in law or altered market conditions; or conduct a separate procurement to secure lease (financing); or to reject the lease option entirely. [27]

Suddenly, with a substantially lower price than expected, NJ TRANSIT began to rethink its financing commitments for the project. Overall capital requirements to build the HBLRT were suddenly much smaller. When the overall cost of the project appeared likely to be much higher, limits on resources led the state to conclude that level payments of $80 million, together with other appropriations, made sense. Included in this approach was the requirement that the contracting provide short-term financing for the project during construction.

But the DBO procurement process had produced a substantially lower price. Resources were now sufficient, maybe, to pay for the actual costs of the project out of current funds, as project costs were incurred. Just before NJ TRANSIT awarded the project to the Raytheon team, the State's Department of the Treasury intervened and sought to pursue a different financing approach that did not include any private financing. The Treasury Department concluded that state-issued bonds, with tax-exempt status, should lower project costs still further.

The New Financing Scheme

Raytheon and NJ TRANSIT agreed to a new financing scheme developed by New Jersey. Grant Anticipation Notes (GANS) were issued as a debt instrument by the state, backed by FTA's commitment to fund the project in the existing Full Funding Grant Agreement of the FTA grant, and further backed by a resolution of the Board of the TTF Authority committing the TTF to apply its annual NJ TRANSIT TTF grant allocation to the project, and further backed by a subsequent NJ TRANSIT Board Resolution binding the Corporation, in the event of a shortfall in FTA grant receipts in any year, to first apply funds received from the annual TTF grants to debt service on the Grant Anticipation Notes. This combination of bonds, grant agreements, and board resolutions was sufficient to sell the GANS in the tax-exempt bond markets.

Questions

1. First, there were five competitors, then three, then two! Was there sufficient competition to make a judgment as to which of the last two teams submitted a price that was too high or too low? How would the existence of a project cost estimate by NJTRANSIT affect your answer here?

Does the gap between the two proposals concern you? Why, and about what? What does this gap indicate about the proposers' perception of the Corporation's RFP? What might NJ TRANSIT learn from the spread in proposal prices?

2. Prepare a short-term financing plan for the project, assuming that 21st Century arranges financing for the design and construction of the project. Assume the following in your analysis.

Total Hard Money Costs for Design and Construction are $724 million. Six percent (6%) of this amount is for design, 94% of this amount is for furnishing the operating equipment and constructing the system.

Design for the project starts in the first quarter of 1996 and construction starts in the 4th Quarter.

The borrowing rate for the Consortium is 10%. The Consortium earns interest at 5% on surplus funds held during the design and construction of the project.

Cash flow for Design over the five quarters beginning in 1996, as a percentage of total design costs, is as follows:

Quarter	1	2	3	4	5
% of Total Design Costs Incurred	30	30	20	10	10

Cash flow for Construction over the seventeen quarters beginning in 1996, as a percentage of total construction costs, is as follows:

Quarter	% of Total Construction Costs Incurred
4	3
5	5

6	6
7	7
8	7
9	8
10	8
11	8
12	8
13	9
14	9
15	9
16	8
17	5

The Corporation's cost for conceptual design is 2% of the estimated total cost of design and construction, incurred in quarter 0.

NJ TRANSIT contributes the funds identified as "Available for DBOM Contractor) in Table 8-2 at the beginning of each year.

At the end of each year, NJ TRANSIT pays the DBOM Contractor an amount sufficient to cover the unpaid balance of design and construction costs incurred in that year plus the contractor's cost of borrowing this amount during the year.

3. Now, prepare a short-term financing plan assuming that the state uses public debt to finance the project's design and construction. State any changes that you make to the financing arrangements now that the design and construction are publicly financed.

   Assume that the interest rate for NJ TRANSIT is 6%.

4. Should NJ TRANSIT proceed with the financing change? If so, should it reimburse Raytheon (or the other proposers) for the cost of including a private financing plan in response to the Corporation's RFP?

5. NJ TRANSIT's capital program is based on a five-year planning horizon and revised annually. Yet, the funds that pay for the capital program come from federal grants, state grants, and cooperative agreements with other agencies that are subject to annual appropriations by state and federal legislatures, administrative decisions by appointed officials in state and federal government, and resolutions passed by public authorities.

How does this dispersed review and approval process affect NJ TRANSIT's ability to plan and implement a sustainable capital program? Can such a system work quickly?

References

Alternatives Analysis/Draft Environmental Impact Statement.

Hudson-Bergen Light Rail Transit System: 21st Century Rail Corporation. 21st Century Rail Corporation. Jersey City, New Jersey, 1999.

Raytheon Engineers & Constructors. "About Raytheon." December 7, 1999. <www.raytheon.com/about/index.htm>

"Bids Hundred of Millions Apart." Engineering News Record July 1, 1996.

"Design-Build Approach Gets Thumbs-up on Transit Jobs." Engineering News Record June 14, 1999.

Duff, J. "Inside New Jersey Transit's $1.1 billion Hudson-Bergen DBOM Light Rail Project." Mass Transit Vol. 24, No. 5, September/October 1998.

Federal Transit Administration. "Report on Funding Levels and Allocation of Funds For Transit New Starts." 1996. United States Department of Transportation, <www.fta.dot.gov/library/money/3jfund/1997/>

"For New Jersey, 'Super-Turnkey' Light Rail Transit." Railway Age September 1996.

"Hudson-Bergen Light Rail." NJ TRANSIT. December 6, 1997. <www.njtransit.state.nj.us/hblrail.htm>

"Itochu Digital Sphere." Itochu Corporation. December 8, 1999. <www.itochu.co.jp>

"Kinkisharyo." Kinki Sharyo. December 8, 1999. <www.kinkisharyo.co.jp>

"Major Capital Project," New Jersey Online. December 7, 1999. <www.nj.com/njtransit/projects.html>

Middleton, W. D. "Innovation: NJ TRANSIT's Diesel LRT Project." Railway Age. 200(2): G1-G6. February 1999.

NJ TRANSIT. Board Item Fact Sheet. NJ TRANSIT, New Jersey, 1996.

NJ TRANSIT. IFB No. 96CT001 Bid Opening: HBLRT DBOM Procurement. NJ TRANSIT, New Jersey, 1996.

NJ TRANSIT. Invention For Bid Package: Contract No. 96CT001. NJ TRANSIT, New Jersey, 1996.

Rubin, D. K. (1999). "Pioneers on the Hudson." Design•Build June 1999.

STV Group. "STV Inc." December 8, 1999. <www.stvinc.com>

Woods, T. J. "NJ TRANSIT's Design Build Operate Maintain (DBOM) Procurement of the Hudson Bergen Light Rail Transit System." Proceeding of the American Public Transit Association (APTA) Rail Conference 1996.

Woods, T. J. "Traditional vs. Innovative Project Delivery Approaches: Cost Management and NJ TRANSIT's Hudson Waterfront Transportation System." Managerial Accounting, Rutgers GSM EMBA Program, Class of 1995.

Notes

[27] Request for Proposals, Book I, 3.6.2 (b)

CHAPTER 9   HIGHWAY 407 ETR (PART 2)

Infrastructure Development Systems IDS-00-T-017

Research Assistant Joe Guerre prepared this case under the supervision of Professor John B. Miller as the basis for class discussion, and not to illustrate either effective or ineffective hand- ling of infrastructure development related issues. Data presented in the case may have been altered to simplify, focus, and to preserve individual confidentiality. The assistance of Miguel Pena of Cintra and Nicolas Villen of Ferrovial are gratefully acknowledged.

Round 1

Alex Bradley's heart sank as the second bid was read aloud at the public meeting. His firm's Cdn$2.75-billion bid for the 99-year concession that would re-privatize Highway 407 ETR had just been beaten by less than five percent.

Alex represented ETR International, a consortium led by Grupo Ferrovial of Madrid and SNC-Lavalin Group of Montreal. ETR International was one of four bidders pre-qualified by the Province of Ontario at the start of the competition for the 407 ETR concession. One had dropped out, unable to reach financing arrangements with their institutional lenders. It had just never occurred to Alan that the two remaining bids would be so close, and worse still, that ETR International's proposal would not be the winner.

But, it wasn't over . . . . yet! The government's Request for Proposal (RFP) required a second round of bidding in the event that the top bids were within five percent of one another. A second competitor was outside the five percent spread that separated the final two competitors. [Note that the Request for Proposals, in this project, culminated in the submission of a "bid", not a "proposal." The term "bid" is typically used when the submission made by a "bidder" to a government is limited to a "price." The term "proposal" is typically used when the submission made by a "proposer" to a government includes factors in addition to "price."] Right after the initial bid opening, each bidder was asked to submit "best and final" offers in a second round of bidding the following week. Alan knew what would occupy his time over the next week --- a searching reexamination of where, how, and if at all, ETR International could cut its price for the concession. The construction cost did not account for the bulk of the bid price, so both competitors were faced with similar problems – how to lower the financing costs paid to sponsoring financial institutions and/or lower the rate of return to prospective equity shareholders in the consortium. With several million dollars already incurred in preparing and submitting the proposal, there would be enormous pressure to lower ETR's price. Known as an "auction" in procurement circles, the two remaining bidders were primarily interested in guessing how their opponent might change its price. Alan wondered what effect the auction environment would have on ETR International's pre-bid assessment of the project.

Background

Highway 407 ETR is a major congestion-relief highway extending 69 kilometers along the north side of metropolitan Toronto. (See, the Highway 407 ETR Part I case study.) Completed in 1998, Highway 407 is the world's first fully electronic multi-lane toll highway – no toll booths and no speed reductions associated with toll transactions. Frequent users save money by purchasing a transponder that attach to the windshield of their vehicle. State of the art technology detects each vehicle as it passes through electronic entrances and exits throughout the highway. Non-registered vehicles are tracked through digital imaging video cameras, that record and process license plate information through a toll-processing center, where tolls are assessed and bills are mailed directly to the owner of the vehicles. The automatic toll collection procedure significantly reduces the delays caused by the gates and booths of traditional toll ways.

Toronto is located along Lake Ontario's north shore. It is the economic center of the Province of Ontario. Before 1998, two east-west routes, Highway 401 and the Gardner Expressway, served the Greater Toronto Area

(GTA). Population and commercial growth in the past three decades overloaded the capacity of the two routes. Congestion stifled the productivity of the GTA and was a major contributor to air-pollution problems. In an effort to address these problems, the Province of Ontario accelerated plans to construct Highway 407 in 1993.

The project was originally structured as a Build-Operate-Transfer contract with a 30-year concession period that was to be privately financed. After two proposals had been submitted, the government split the proposals into two parts, and selected the road portion from one and the tolling portion from the other. The project was divided into two separate contracts, one for the Design-Build of the road, and the other for the supply and operation of the toll system. The Province of Ontario financed both contracts with government funds, investing Cdn$1.5-billion (Cdn$1.00 = US$0.68) in design, construction, and initial operation of the project. The original project was obtained through a combination of two delivery methods: the first for design and construction of the physical roadway (a Design-Build contract in Quadrant IV of the quadrant framework); and the second for furnishing, installing, and operating the electronic toll system on the project (a directly financed Design-Build-Operate contract in Quadrant I). The government obtained prices for each of these components and used these prices as a floor for establishing fixed price commitments for the delivery of the entire project.

The Re-Privatization of Hwy 407 ETR

After another election cycle (and a change in government), officials continued to explore the feasibility of attracting private investors to the project. On February 20, 1998, the Province announced that it would conduct a competition to "lease" the 407 ETR project to private operator/investor. In return for the right to collect tolls, the winning consortium would assume responsibility for (a) operating and maintaining the project throughout a 99-year concession period, and (b) for constructing, operating, and maintaining two extensions to the road totaling 39 km. The two extensions, one to the west and one to the east, had been planned during the original design process, and were important to the government for long-term congestion relief.

Exhibit 9-1 includes three maps showing the layout of 407 ETR and the extensions. The "East Partial" is the portion of the preliminary design completed at the time the RFP was issued. The white-highlighted area to the east of the "East Partial" is the "East Complete" section, for which no design work had been completed by the government at the time the RPF was issued. The "East Complete" section was later withdrawn from the scope of the competition during the procurement process because the government concluded that uncertainty in traffic forecasts for this area introduced an excessive amount of risk into the project.

Exhibit 9-1   *Layout of the Extended Highway 407 ETR Project*

The delivery and finance method chosen by the province was a very-long term Design-Build-Finance-Operate contract that falls in Quadrant I of the quadrant framework. The government concluded that the leasing of Highway 407 had three distinct advantages. First of all, Ontario would receive a large sum of money up front, which could be used to pay off the debt incurred through the original DB contract, and apply the remainder toward other essential infrastructure projects. Secondly, the province would transfer the long-term risk of making a return on their Cdn$1.5-billion investment to the new lessee. And finally, Ontario and the GTA would enjoy the benefits of an additional 39-km of fully automatic toll-road at no additional costs.

Figure 9-1 maps the project delivery and finance strategy for the 407 ETR projects into the quadrant framework.

The RFP

The Province issued its RFP for the 407 ETR concession on December 12, 1998, and included a clearly defined scope of work, a detailed statement establishing the respective obligations of the government and the successful concessionaire (a clear risk allocation), and complete financial information showing historical revenues and costs associated with the project. To create a transparent procurement process, the Province accomplished a number of specific tasks before the RFP was issued. The Province:

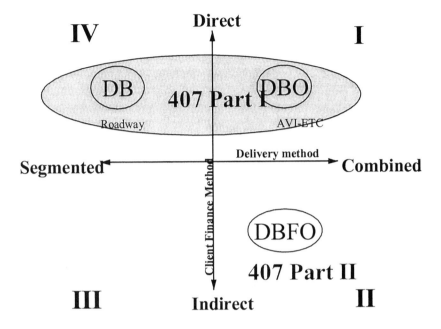

Figure 9-1   *The Highway 407 Project Delivery Strategy*

- provided a detailed, yet simple, description of the toll structure it would require the successful concessionaire to implement;

- provided a detailed statement of the work the successful concessionaire would be required to perform over the term of the concession to maintain and operate the existing facility, and to design, build, operate, and maintain extensions to the facility;

- established a process for bidders to have complete access to existing information (both financial and technical) relating to the road;

- included the form of legal agreement the Province would require the successful bidder to execute upon contract award; and

- clearly stated the Province's single bid evaluation criterion, highest price paid to the Province for award of the concession.

Profit Projection Data: The RFP included information that enabled bidders to make their own evaluation of the profit potential of the project by disclosing revenues generated by the facility in the nine months preceding the RFP. Figure 9-2 shows historical revenues for the period preceding the RFP. The Province also paid for several independent traffic studies that used historical traffic counts and revenues to project future traffic counts and revenues under the concession arrangement. Bidders were given unlimited access to each of the studies as a starting point for their own evaluation of the profit potential of the project. Figure 9-3 shows the range of traffic projections provided by the government to bidders.

OVERVIEW OF TOLL REVENUES

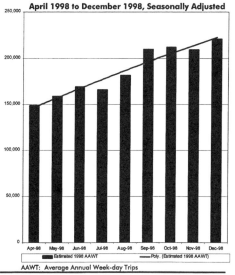

...the Highway has been running at approximately $11MM per month in toll revenues.

(NB) NESBITT BURNS

Figure 9-2   *Actual Revenues Prior to the RFP*

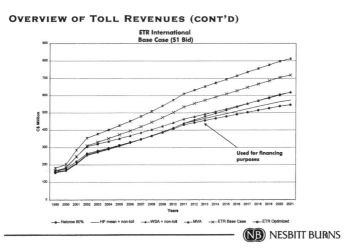

OVERVIEW OF TOLL REVENUES (CONT'D)

Figure 9-3   *Independent Projections of Concession Revenues*

Finally, the Province carefully established its requirements for the toll structure on the road prior to the issuance of the RFP. This toll structure was included in the RFP to be used by the competitors in establishing their proposed concession price. The toll structure incorporated the Province's stated goal of moving 9,000 vehicles per hour across the roadway at peak hours – the congestion capacity of the highway established by the Province. The toll structure established a number of incentives for the concessionaire to focus on high quality performance and congestion relief throughout the concession period. Key elements of the toll structure were:

Regulated Rates Below Congestions Capacity

1. The maximum toll for <u>cars</u> for the first year was set at Cdn$0.10 per km.

2. During the first fifteen years of operation, the toll rate could only increase annually by an amount equal to the sum of the inflation plus 2%, but tolls could not exceed Cdn$0.13 per km in the first 15 years. The inflation rate was established by an independent index referenced in the RFP.

3. During years 16 through 99 of the concession, the toll could only increase annually at the rate of inflation.

4. Toll rates for trucks could be up to twice the rate for cars. Toll rates for multiple-unit trucks could be up to three times the rate for cars.

5. Vehicles not fitted with a transponder could be charged an additional service fee of Cdn$1.00 per trip.

Incentives/Penalties above the Congestion Capacity

6. If peak hour flows (a defined term in the RFP) remained above 9,000 vehicles per hour (VPH), all rate restrictions would be dropped. In these circumstances, the Province's congestion goal would be fulfilled, and the concessionaire would be allowed to adjust fares (upward) to establish fares that maintained traffic on the road at the congestion level.

7. If peak hour flows fell below 9,000 VPH during period where toll rates charged by the concessionaire were higher than rate limits described in paragraphs 1 through 5 above, the concessionaire would be penalized according to the following equation:

Two times (Congestion Rate minus Regulated Rate) times (9,000 minus Actual VPH)

By removing toll rate restrictions above the congestion traffic volume (9,000 VPH), the Province concluded that the quality of service on 407 ETR would be maintained at or near the congestion traffic volume through market adjustments by the concessionaire in the tolls. The Province's toll rate structure encouraged the concessionaire to find and maintain this market price, since prices that were either too high or too low are sub-optimal to the concessionaire. Too high a toll pushes traffic away to other routes and traffic volumes below the 9,000 VPH congestion rate. Too low a toll attracts too many users, and would tend to reduce the speed of traffic across the road (another reduction in total receipts to the concessionaire). This toll structure also encourages the concessionaire to keep the road fully open and in good repair and so as not to lose revenue. For example, the toll rate structure encourages off peak repairs, quick attention to disabled vehicles, and to other activities that minimize accidents or disruptions, e.g. constant attention to signs, road marking, cleaning, drainage, plowing, sanding.

Transparent Access to the Same Information

Ontario took several steps in the RFP to assure that all bidders received consistent information. First, the bidders were required to submit all questions in writing. Questions were answered in writing, and circulated to all bidders. Approximately 500 questions were answered over the three month bidding process. Secondly, the Province named a single employee as the contact point for all communications between government officials and bidders. Several pre-bid conferences were held during the bidding process, usually attended by all bidders. When one-on-one meetings were required between a single bidder and a government official, an independent third party advisor confirmed that all issues were uniformly addressed across all bidders. The Province also established three data rooms to assist the bidders. The rooms contained a large collection of preliminary plans, as-built drawings, specifications, and reports. The RFP required the bidders to make appointments in order to visit the rooms and review the material.

Establishing a Floor for the Content of Proposals Through A Standard Concession Agreement

Prior Review and Comment on the RFP's Standard Agreement

Early in the procurement process, before the RFP was issued, drafts of the concession agreement were distributed to all parties interested in bidding. The draft agreement contained a detailed statement of the scope of work for the project, and the specific requirements that would be imposed on the successful concessionaire. After time for review, the Province received initial comments and suggestions as to how the concession agreement could be improved. The comments were collected, reviewed, and the agreement was modified. The redraft was again circulated to potential bidders, with encouragement to submit further comments. After two review cycles, the final agreement was issued to all the bidders. The RFP required that the all bids acknowledge that the terms of the concession agreement would be accepted, without amendments, upon award of the concession to the successful bidder.

Pre-Qualification of Bidders

The Province conducted a vigorous pre-qualification process, which was used to establish the technical and financial capabilities of the various consortiums to design, build, finance, operated, and maintain the project over the concession period. The pre-qualification process confirmed the financial capabilities of the bidders, and confirmed that each consortium had the necessary experience to successfully complete the obligations under the concession agreement.

Evaluation Based Solely on Price

The Province established a single bid evaluation criterion for award of the contract: price. Bids were submitted, not proposals, since submittal of financial assumptions, traffic projections, and pricing calculations were not required. Only the final price was reviewed. Instead of awarding the concession contract to the *lowest* responsive responsible bidder, the Highway 407 concession was awarded to the *highest* responsive responsible bidder. The evaluation process was simple, and allowed the Province to award and finalize the concession contract within a matter of days of the bid opening. Because of the simplicity of the evaluation process, bidders had to be in a position, within several days of bid submittal, to execute the standard concession contract with bank financing already in place and committed. This was a significant challenge to bidders, who had to finalize all supporting financial arrangements and agreements prior to the submittal of bids. These difficulties proved too much for one of the four pre-qualified bidders, who out of the competition. Three bids were submitted on March 30, 1999.

The government's evaluation process in the first round was completed in a matter of days, during which the Government had to decide whether they wanted to execute a concession agreement for the Base project or for the Base plus one of the three alternatives which extended the project to the East. After the decision was made to stick with the Base project and the Government received best and final offers, the evaluation of the second round of bids took only a few hours and an awarded was immediately announced.

While a price-only competition assured transparency in the assessment of bids, the pre-qualification process was critical to the process. Because of the government's decision to evaluate only on price, the inclusion of the standard concession agreement in the RFP was a second critical factor in the success of the procurement. The existence of the concession agreement ensured that each proposal met or exceeded the Province's financial, technical, and environmental requirements.

Standard & Poor's Risk Assessment

ETR International obtained a letter from Standard & Poor's that provided a preliminary evaluation of the project, which is reproduced as Exhibit 9-2.

Exhibit 9-2 Standard & Poor's Evaluation Letter

The letter identified the following risks:

- Legal and business risks associated with an untested procurement process.
- Construction risks associated with the extensions.
- Pricing risks and performance risks associated with the long-term operation and maintenance obligations.
- Uncertain future traffic volumes are directly related to the ability to service the project debt.

The letter also describes the strengths of the project that served to mitigate the above risks:

- The project is located in a region that has a stable political, legal, and economic environment.

- The owner of the project is sophisticated as to technical and economic matters relating to the project.

- A long concession period.

- The successful current operational performance of the existing facility and tolling equipment.

- The facility's proven ability to produce revenues combined with substantial evidence that traffic volumes will continue to grow.

- The technical requirements needed for the construction of the extensions are manageable.

The project received a single-'A' preliminary rating by Standard & Poor's.

ETR International's First Bid Price

ETR International's initial bid price was Cdn$2,750,000,000. The bid was based on a project-financing plan that consisted of 17% private equity and 83% debt. The equity portion of the plan, totaling Cdn$500-million, equaled the estimated cost of design and construction for the two extensions to the 407 ETR project. The debt structure consisted of a mixture of senior bonds, subordinate bonds, and convertible bonds, which were to be sold in both the Canadian and US markets. Table 9-1 summarizes the proposed refinancing schedule for the bank bridge facility and the sponsors' subordinated debt.

Note that convertible bonds were considered by the Consortium as equity for all practical purposes. Restrictions established in the RFP encouraged competitors not to treat these bonds as straight equity.

Exhibit 9-4 contains a simplified version of ETR's complete financial analysis supporting their initial bid, with revenue and expense projections included. In sheet titled "Calculations", the row entitled "Revenue from

Operations" includes all O&M expenses, toll revenues, taxes, and tax benefits. At the bid price first submitted by ETR, these calculations indicate a NPV of Cdn$39-million for the project over the first fifty-five years of the ninety-nine year concession, the year all project debt will to be retired. The NPV of the last 44 years of the contract is not analyzed.

Senior Debt Financing		Amount (000)s	Issue Date	Date Due	Term (years)	Amortization (years)	Spread (bps)	Coupon Rate
Tranche 1	C$ 10 Yr Bullet	400,000	6/30/99	6/30/98	99	0	90	6.40%
Tranche 2	C$ 30 Yr Bullet	300,000	6/30/99	6/30/98	99	0	90	6.40%
Tranche 3	C$ Amortizer	300,000	6/30/99	6/30/39	40	35	90	6.40%
Tranche 4	US$ 7 Yr Bullet	337,500	12/30/99	12/30/98	99	0	115	6.65%
Tranche 5	US$ 10 Yr Bullet	337,500	12/30/99	12/30/98	99	0	115	6.65%
Tranche 6	C$ Amortizer	525,000	6/30/00	6/30/40	40	35	90	6.40%
Tranche 7	C$ 7 Yr Bullet	150,000	6/30/00	6/30/99	99	0	90	6.40%
Tranche 8	C$ 10 Yr Bullet	150,000	6/30/00	6/30/99	99	0	90	6.40%
Subordinated Debt Financing								
Tranche 1	C$ Amortizer	170,000	12/30/01	12/30/21	20	15	150	6.700%
Tranche 2	C$ Amortizer	913,000	12/30/01	12/30/21	20	15	150	6.700%

Table 9-1   *ETR International's Proposed Refinancing Schedule*

Exhibit 9-3   *ETR Financial Analysis*

Questions

1. Whether the competing bid was higher or lower than ETR International's bid of Cdn$2,951,400,000 was unknown – just that the two bids were within 5% of each other. Five days before "Best and Final Offers" are due to the Province, the Presidents of the member companies of ETR International asked Alex to explore three different scenarios in a report to the Board of Directors. Each scenario involves adjusting the financial structure of ETR's bid. The three scenarios are:

Scenario 1: increase ETR's bid by one-half of one percent, or $14.757 million dollars.

Scenario 2: increase ETR's bid by one percent, or $29.514 million dollars.

Scenario 3: increase ETR's bid by one and one-half percent, or $44.271 million dollars.

Using the spreadsheet in Exhibit 9-1 as a guide, review the financial structure of ETR's proposal. **Without making changes to the spreadsheet or performing any quantitative calculations, answer the following questions qualitatively.**

a.   How can ETR International change elements of the project's financial structure to increase its bid and what impacts will these changes have?

Review rate assumptions and projections for revenues and expenses. Consider whether changes in rates (for example changes to interest rates, return on equity, discount rates, etc.) or cash flow projections (for example changes to construction costs, tolls, operations & maintenance, etc.) might produce changes in the financial analysis that approach the desired results in Scenarios 1, 2, or 3. Prepare your recommendations to the ETR Board summarizing how to alter the project's structure to increase ETR's bid. Also, describe the financial impact of your proposal on each of the parties in the consortium. Think about the financial stake each party has and how it might change based on your recommendations.

b. Are your recommendations principled changes in the project's structure or a shift in previous assumptions to back into high numbers?

c. What do you recommend ETR International submit as its Best and Final Offer? Why?

2. Best and Final Offers are frequently used to allow proposers (not bidders) to revise both the technical and financial sections of their proposal to increase the value of their submission to the client. Were best and final offers appropriate in this situation? Why? Why not?

CHAPTER 10   INDIANAPOLIS WASTEWATER TREATMENT WORKS

Infrastructure Development Systems IDS-97-W-101

Research Assistant Maia A. Hansen prepared this case under the supervision of Professor John B. Miller as the basis for class discussion, and not to illustrate either effective or ineffective handling of infrastructure development related issues. Data presented in the case has been altered to simplify, focus, and to preserve individual confidentiality. Names of employees and contracting firms have also been changed. The assistance of The Honorable Stephen Goldsmith, Mayor of the City of Indianapolis, and Robert Hawkinson III in the preparation of this case is gratefully acknowledged.

Mayor's Office, Indianapolis, September 1993

Steven Goldsmith enters the crowded executive conference room and takes a seat at the head of the table. Surrounding him are representatives from the major municipal departments in Indianapolis, including the Police Department, Fire Department, Community Affairs, Personnel Department, Treasurer, Procurement, Social Services, Parks and Recreation, and Public Works.

Mayor: "Good morning, everyone. I understand we've got a lot to discuss today, including some important decisions to make about key contracting and funding issues. We have a number of proposed special projects for the coming year. I hope we can reach a consensus today on how we should allocate our limited resources. I know everyone is busy right now. So let's get started. Russell, why don't you start with your presentation on our contracting study of the wastewater treatment plants. I've had so many people calling me about this and I don't really know what to tell them."

Russell gets up to speak...

Background

Russell James grew up on a small family farm in Southern Indiana. As a bright high school student and star athlete, he earned a scholarship to Purdue University, where he majored in Civil Engineering and became the captain of the swim team. When he graduated, he took a number of jobs in engineering and environmental compliance firms in the Midwest, primarily in Chicago. Eventually, he returned to Indiana and to Indianapolis, where he wanted to help the new city administration reexamine the way the municipal infrastructure was operated, improving both environmental quality and the city's budget situation. For over a year now, Russell was in charge of the Water and Sewer Division, within the Public Works Department. He oversaw a staff of 600 including all operational and administrative sections. Aside from day to day operations, his role had expanded to be internal consultant to a mayor particularly interested in infrastructure issues. Currently, Russell was heading up the highly publicized contracted operations of the city's large wastewater treatment plants.

Mayor Goldsmith had held office as the mayor of Indianapolis, IN, the twelfth largest city in America, since November 1991. He was well known in national politics and often cited by the media for his innovative approaches to managing government. National interest in debt reduction and streamlining government operations was growing and Indianapolis was often hailed as a model for others to emulate. Goldsmith's efforts as mayor were on reducing government spending where possible, cutting the city's bureaucracy, limiting taxes, and reducing counter-productive regulations, while improving the quality of services to residents with more innovative, responsive programs.

History of Indianapolis

A crossroads for travel and transportation for Indiana as well as the entire nation, Indianapolis is a major industrial and wholesale-retail center and a grain and livestock market. The city's industries produce automobile and airplane engines, electronic and electrical equipment, pharmaceuticals, chemicals, furniture, and machinery.

Indianapolis was founded in 1820 and became the state capital officially in 1825. Its growth was spurred by the opening of the National Road in 1827 and later by the development of the railroad industry. Indianapolis was incorporated as a city in 1847. In the late 1800s, natural gas was introduced and the new cheap fuel attracted much industry to the city. From these beginnings, the city has grown to 352 square miles and a population of over 700,000 within the city limits and 1.4 million people in the nine-county metropolitan area. The largest employer is the local government with 62,700 employees, followed by state and federal government with a total of 47,000. Other major employers are Eli Lilly and Company (7,500 employees), Marsh Supermarkets (7,000), St. Vincent Hospitals (6,000), Delphi Interior & Lighting (4,425), and Allison Transmissions/ GMC (4,200).

The Per Capita Personal Income for the metropolitan district is $22,019 and the unemployment rate is 4.0%. A percentage breakdown of employment shows that 25.1% of working people are in the service industry, 19.1% in retail, and 16.4% in manufacturing. There is a state retail tax of 5% and a state adjusted income tax of 3.4%. The government of Indianapolis and surrounding Marion County were consolidated in 1970 and are headed by a mayor and city-county council.

History of the Wastewater Treatment Plants

The City of Indianapolis, Department of Public Works, operates two wastewater treatment and solids handling facilities. These facilities are know collectively as the Advanced Wastewater Treatment (AWT) facilities, and consist of the Belmont Advanced Wastewater Treatment Plant, the Belmont Solids Handling Facilities and the Southport Advanced Wastewater Treatment Plant. The Belmont and Southport AWT plants utilize advanced treatment processes with a total average treatment capacity of 245 MGD (125 mgd each). The facilities include preliminary treatment, primary clarification, biological treatment via bio-roughing and oxygen nitrification, followed by secondary clarification, effluent filtration and ozone disinfection prior to effluent discharge to the White River. The plants employ 433 people.

The two plants were first opened in 1982 together with other facilities constructed between 1980 and 1990. Total cost of the system was $372 million, of which approximately $284 million was provided by federal EPA and state grants pursuant to the Clean Water Act of 1972. (See Table 10-1). The remaining funding from the city was financed through municipal bonds and supported by user fees.

	Year Built	EPA/State Share	Locally Funded	Total Cost
Belmont WWTP	1982	122,070,145	35,059,423	157,129,568
Southport WWTP	1982	85,964,845	23,888,728	109,853,573
Sludge Management Facilities I & II	1990	52,853,775	24,179,566	77,033,341
East Marion County Regional Interceptor	1980	8,827,003	1,557,706	10,384,709
South Marion County Regional Interceptor	1982	14,780,851	2,608,386	17,389,237
Total		284,496,619	87,293,809	371,790,428

Table 10-1   Cost of Facilities

The value of the AWT facilities is difficult to pinpoint, because calculations vary greatly depending on assumptions made. The plants are worth $79 million where rates and revenues are estimated for a not-for-profit company or continued city ownership. The AWT facilities are valued at $247 million, however, using the rate base valuation methods that simulate the return a for-profit, regulated company would be allowed to earn. Reproduction cost less depreciation produces yet another estimate: at $307 million. The values for the facilities are based on original cost data as well as the revenue requirements developed in studying current operations and capitalized income streams. The facility values also include materials and supplies, working capital, and other assets used for operation. Exhibit 10-1 on the companion CD summarizes cash expenses for the system.

Exhibit 10-1   Cash Base Analysis

The city currently has outstanding debt of $139 million for the Sanitary General Fund, of which approximately half was attributed to the AWT facilities. Annual debt service was estimated at $9.7 million for the base year (1993) and projected to increase as shown in Table 2: Cash Basis Analysis of AWT Revenue Requirements. The debt service is financed through property tax bills paid by all taxpayers within Marion County. Therefore, the actual cost of sewer service is understated in the user fees because of the property tax subsidy. The current sewer rate includes a minimum monthly charge for non-industrial customers and industrial customers of $5.43 and $5.59, respectively. This monthly charge consists of a minimum usage fee of 3,000 gallons and a connection fee, which recovers a portion of the costs of un-metered flows and infiltration. The charge for each additional 1,000 gallons is $1.13 for non-industrial customers and $1.19 for industrial. The current rates do not represent the cost of operating the waste water system, however. Current bills are approximately 35% less than cost. Under private ownership, user bills would be an additional 30% to 50% higher than under City ownership, due to the debt financing currently paid from property tax revenues.

Based on present estimates by the environmental consulting firm, renovating the facilities to peak operating capabilities will cost the AWT $300,000 in immediate repair (as shown in Table 10-2), $8,363,000 in capital improvements over the next three years and a total of $142,437,653 for the Capital Improvement Program over 20 years. The capital program is assumed to be funded in the future with 20% pay-as-you-go and debt financing for the remaining 80%.

Initial Study of Operating Alternatives

Russell had been at his new job for only a few months, when the issue of the competition for the operation of the wastewater treatment plants was raised. The option of competing the treatment plants in the public and private sectors soon became an important issue in the office. Given the focus of the mayor and other city leaders, Russell felt it might be the right time to study different types of ownership and/or operation of the AWT.

Facility	Total Estimated Cost of Projects	Average Annual Cost for 3 Year Projects
Belmont AWT		$1,228,000
Headworks	$900,000	
Primary	$664,000	
Bioroughing	$1,065,000	
Oxygen Nitrification	$950,000	
Effluent Filters	$400,000	
Ozone Disinfection	$50,000	
Miscellaneous	$385,000	
Belmont Solids Processing		$446,667
Sludge Thickening	$100,000	
Dewatering / Incineration	$980,000	
Misc.	$200,000	
Southport AWT		$1,113,000
Headworks	$340,000	
Primary	$0	
Bioroughing	$985,000	
Oxygen Nitrification	$1,539,000	
Effluent Filters	$650,000	
Ozone Disinfection	$70,000	
Miscellaneous	$505,000	

Including: 1. Projects that will be required to meet future conditions such as increased flows and loadings and new regulatory requirements; and

2. Projects that feature new technologies for increased operational efficiencies

Consultants' Recommended Annual Budget for each facility from 1996 forward
     Belmont $1,000,000
     Belmont Solids* $ 700,000
     Southport AWT $1,000,000
Includes $200,000 for new solids processing room

Table 10-2   *Summary of Three Year Projection for Required Capital Improvements*

Since the financial, tax, and technical aspects of the AWT were quite complicated, and Russell's administrative staff was small, he hired two consulting firms to study the plant. One of the top national accounting and consulting firms started its study late in 1992 and was soon joined by an equally large environmental design firm. Their task was three fold: (a) to establish a baseline and 20 year projection for the revenues and expenditures of the facilities; (b) to determine the value of existing assets; and (c) to define alternatives to generate new revenue for wastewater capital improvements such as sale, lease, public/private venture or expanded service. The consultants considered several operation / ownership models including the following:

1. Enhanced City Ownership and Operation

2. Contract Operations with Continued City Ownership

3. Establishment of an Independent Authority

4. Establishment of a Not-for-Profit Corporation

5. Sale or Lease with an Operating Contract for Service

6. Establishment of an Investor-Owned Utility

In the spring of 1993, the accounting firm published its report.

Russell flipped open the report and went directly to the last section: Findings and Recommended Assessment Action Plan. (Summary as Table 10-3) After reviewing the report, Russell concluded that the three most important issues were: (1) Assessing user charges, (2) Obtaining bids for contract operations of the AWT facilities, and (3) Improving Capital Investment.

As he continued through the report, Russell found the consultant's assessment that identified contract operations with continued City ownership as having the most favorable net present value. This alternative would have minimum impact on user charges, with a potential decrease in charges. Additionally, it was expected to have good vendor interest and little legal and regulatory challenges. Over time, the private operators could be expected to reduce costs by 5%. To quote the report: "Because the city had made advances in improving efficiency, a private operator would have limited opportunity to reduce costs initially, and would be reluctant to risk operation of this large, complex facility or permit requirements. However, over time (three years) we believe they would gain the knowledge of operations and confidence to operate with sewer staff and reduce personnel costs by five percent in 1997."

Recommendation	Rationale	Steps
Financial		
Update User Charges	Current rates do not cover expenses without help from property taxes	Determine revenue requirements; revise user charge methodology
Enhance enforcement of highly contaminated waste	Current program results in a loss of $1-2 million every year in revenue. Plant needs upgrading	Identify more comprehensive list of industrial waste generators; Enhance monitoring; Charge for excess levels
Operations		
Enhance enforcement of highly contaminated waste	See above	
Redirect overhead costs not required for AWT facilities	Revenue requirements exclude indirect/overhead costs not required	Identify overhead/indirect costs not required by AWT facilities
Use existing reserves for capital improvements	funds are in place, plant needs upgrades	determine acceptable draw-down level; Fund select projects
Revenue Alternatives		
Obtain bids for contract operations of AWT	Potential for Savings; Estimates difficult to determine without firm bids	Enact rate increase; Develop performance measures; Obtain bids
Consider establishment of an Authority of AWT (following successful contract operations)	Preferable form of ownership; Allows for flow of sales proceeds to City	Await results of contract operations
Policy		
Evaluate policy of using property tax revenues to fund AWT facilities debt service	Tax payers not using AWT should not have to fund operations; Property tax payments do not reflect proportionate use of AWT facilities	Obtain legal opinion

Table 10-3   *Recommended Action Plan*

As a result of the study, Russell and the mayor's staff concluded that, of all the alternatives studied, a contract operation strategy with continued City ownership would produce the best option for the City's investment. They decided to advertise the operations contract with the maximum five years allowable under Federal Law. (See Exhibit 10-2 on companion CD)

Exhibit 10-2   Summary of Federal Laws

The timeline of the competition is shown below. The primary evaluation criteria for each bidder were: financial strength of the proposal based on Net Present Value, technical abilities, experience, and specific additional proposals. A complete copy of the RFP is attached as a reference exhibit.

Timeline of Procurement Process

June 1, 1993	Issue RFQ
July 15, 1993	Issue RFP
July 26, 1993	Facility Tours
Aug. 10, 1993	Pre-proposal Meeting
Aug. 27, 1993	Proposals due to Review Committee
Sept. 30, 1993	Expected government Decision
Dec. 20, 1993	Scheduled O & M agreement signed between parties

Division of Responsibilities

City	**Contracted Operator**
Ownership of land, blds, vehicles	Operation - including matls, equip, utilities(at or above previous performance. stds)
Monthly / Annual mtg w. contractors	Maintenance of facil / equip
Fund Capital Improvements	Employee payment, benefits, etc
Collection of sewage fees?	Training, Safety
Customer Service	Monthly recording/reporting, lab analysis
Contract Oversight	Transport of sludge
Fund changes in scope	Monthly and Annual Expenditure Reports

Co-insurance for personal inj/prop damage	Monthly review with city of ops, etc
Turns-over fac. & Equip in good order	Insurance - Indemnification to city
	Performance Bond - One Year
	Management of Capital Improvements
	Co-insurance for personal inj / prop damage
	No subcontracting w/o prior written consent
Prepares permits for routine renewal, etc	Prepares permits for Process Change
Process reports for local, state, fed authorities	Assists city in processing reports
	Contribute 5% of pre-tax profits to support economic development initiatives

The Contract Selection Board Meeting

The city prepared and issued an RFP, and received five proposals. Approximately nine months after hiring the first consultants, Russell found himself analyzing the final proposals for the project. His staff and the procurement office had whittled the competition down to three organizations, and now it was time to make the final decision. To prepare for making a decision, he had hired the original consultants to analyze the financial, technical and community implications of the final proposals. In addition, a member of his own staff had prepared a comparison of the three groups. With this information, he brought together the advisors for the contract selection board and prepared for a long meeting. The mayor had been asking for his recommendation for a week and he needed to have a solid presentation ready for the executive board meeting on Friday.

Public Works Conference Room: Tuesday, 9 a.m.

Russell: "Good morning everyone. Thank you for taking time out of your busy schedules to help my department make this important contracting decision. Our selection will receive strict scrutiny from the local citizens and the national press. Don't change your decisions because of this, but be aware that we need to have all of our facts straight when we present this to the mayor. We will be meeting this morning to present the studies and again tomorrow to vote on the final selection."

"Why don't I quickly introduce everyone. To my right, we have Jane Messinger, the head of procurement. Her office has been extremely busy tracking this competition for the last three months. Thanks, Jane. Next to Jane is Carolyn Miller, the lead consultant for Environmental Solutions. Beside her is Paul Wilson, a CPA from Ellery and White. Next is Vince Brown, my assistant operations officer. Beside Vince is the city lawyer, Henry Unger. And finally, we have Katherine O'Connor, a colleague from the Water Conservancy District and president of the local American Society of Civil Engineers chapter."

"As most of you know from the memo I distributed at the end of last week, we have narrowed the competition to three organizations (See Exhibit 10-3 on companion CD). All three meet the credentials of operating our plants satisfactorily. From these three, we need to find the best group - comparing all of them against the strict financial and technical criteria that were published in our RFP for these services. One option is to cancel the solicitation and award no contract, but we would like to avoid repercussions of a cancellation on future procurements in the city, if we can."

Exhibit 10-3   *Summary of Top Three Competitors*

"I'd like to start by having the three reports presented by the respective writers. Carolyn, would you like to start?"

Carolyn Miller, Environmental Solutions:

"Thanks Russell. Good morning everyone. As you know, my part of the analysis evaluates primarily the technical aspects of each bid. We found that, in general, all of the groups were competent and could do the job. Please take a copy of the overheads that I am passing around, so that you can follow me as I talk. (See Exhibit 10-4 on the companion CD). In our analysis, we concentrate on companies A, B, and C. Initially there were five groups competing as shown in the first summary. The three we will discuss today presented the most complete and responsive proposals and were the most attractive to the city."

Exhibit 10-4   *Technical Proposal Review*

"In our opinion, the key areas of technical concern are these: (a) significant reductions in staffing proposed by each of the contractor, (b) the contractor's ability to meet current and future permit requirements, and (c) protection of the City's investment in the facilities. "

"My evaluation criteria was based on staffing, protection of investment, ability to meet permit requirements, and other risks, including adequate proposed capital programs, safety and contractor / owner relationships. We conducted six site visits, including two facilities currently operated by company A, three operated by B, and the AWT plant currently operated by company C, the employee run bidder."

For each company, we graded the risk of every aspect of their operations. Based on this qualitative analysis of each area, we developed the risk assessment table shown in Exhibit 10-5 on the companion CD. The following matrix table summarizes the results of that study:

Exhibit 10-5   *Technical Risk Assessment*

Company	A	B	C
High Risk Elements	12	14	-
Risks Equivalent to Current Operations	12	8	27
Low Risk Elements	3	5	-

Table 10-4   *Risk Matrix*

Based on the maintenance cost breakdowns provided, Company A's proposed budget for preventive/predictive maintenance presents a concern as indicated by the following budget comparison:

	Preventive / Predictive Maintenance Budget
A	$284,287
B	$843,111
C	$2,000,000

"On the issue of staffing, we feel that both A and B can operate and maintain the facilities in an acceptable manner, as evidenced by their current operations in other locations. A's plan involves the most risk because they propose significantly lower staff levels than the others. B's transition is also risky, though, because they will make immediate staff reductions, rather than gradual changes."

"As for maintaining the facility and meeting permits, both A and B have similar capabilities and risk assessment. They both plan to make changes to the existing plant, but we think we can evaluate their performance with the aid of appropriate contract provisions (separate maintenance budgets, specific definitions of maintenance vs. capital expenditures, etc.). Both A and B propose significant process modifications. These changes present high-risk elements compared to the employee's (Company C's) proposal. New permit requirements due to changes in the plant processes pose the risk of increased O&M costs. Capital improvements may require a contract amendment to provide additional compensation to A or B. The C proposal presents only moderate risks in this regard,"

"My conclusion, based on these broad factors, is to recommend that the award be made to either group A or B."

Russell: "Thanks Carolyn. OK, Paul, could you do the same for your analysis of the financial and community issues:"

Paul, E&W Accounting:

"Sure. I have similar handouts for the group...."

"The City of Indianapolis requested E & W to comment on the net present value calculations of costs proposed by each of the competitors. In particular, we were tasked to advise the city on the reasonableness of the net present value methodology and the calculations used to compare the cost proposals. In general, we found that the calculations were done correctly. I've attached the respective spreadsheets to your handouts, for your comparison (See Exhibits 10-6 through 10-9 on the companion CD). From these calculations you will understand the bottom-line costs of each group as well as the specific cost allocations in each group."

Exhibit 10-6	NPV of Company A
Exhibit 10-7	NPV of Company B
Exhibit 10-8	NPV of Company C
Exhibit 10-9	NPV of 1994 Budget

Russell: "Before closing for the day, I'd like to give Vince a chance to discuss his findings from the point of view of the Public Works Department.

Vince: Public Works:

"I had the benefit of seeing both reports before completing my analysis, so I've tried to develop my slides with the same structure as the previous consultants."

"In general, I think the city should enter into a contract with a private firm as soon as practical. Private contract management will preserve the outstanding environmental record of Indianapolis and will save $12 million in operating expenses in 1994 and at least $65 million in operating and capital costs over the five-year term of the contract. After conducting five site visits to currently operated sites, meeting with present customers, visiting operational headquarters, and conducting financial and environmental analysis, we have determined that all three organizations could do the job, but we are leaning toward selecting A or B.

"As seen in Paul's calculations, all three companies provide cost savings to the city. Although company C's bid provides fewer cost savings, we should note that the existing operation of the AWT facilities by the AWT management group is excellent. Moreover, the employees' proposal yields efficiency cost savings beyond the 1994 Budget. Inherent in the proposals by A and B are transition matters that must be managed effectively. In this case, our two major transition issues are:

1. Process changes that must be implemented without impacting permit compliance

2. Substantial reduction in AWT employees from the present level.

I should also note that company B has established its headquarters in Indianapolis and will be developing an international training and research center here.

Russell: Thanks Vince. I think we need to take a break and let these details sink in before we continue with our debate. Let's stop for the day. I'll see all of you again tomorrow at 9 a.m. Thank you.

Russell and Katherine, a friend from his undergraduate days at Purdue, leave to buy sandwiches at the corner deli.

Katherine: "Russell, this decision is pretty complicated. I'm glad I'm the unbiased outside party on this project. Why don't we grab lunch and make some back of the envelope calculations while we eat. You've got a lot of work ahead of you if you want to be ready for that meeting on Friday."

During and after lunch, Russell and Katherine put together some rough numbers so that they would be able to discuss the alternatives intelligently the next day.

Questions

1. Compare the potential cost savings for each scenario. Consider additional factors in calculating the NPVs. Weigh the additional risks.

2. List in priority order what issues you think the Mayor would want to be briefed on.

3. What additional questions will you ask the consultants and your staff tomorrow? What information would you still like to receive?

4. Where does the original design and construction of the AWT fit in the quadrant analytical framework?

5. Where does each of the alternatives originally under consideration fit in the framework?

6. What savings might be achieved through contract operations?

7. Evaluate the criteria by which the competitors were judged. How would you approach future competitions? What challenges did the city face in trying this competition for the first time? What procurement/delivery style do you recommend in the long term?

References

Request for Proposals. AWT Facilities. Indianapolis, Indiana, July 1993

CHAPTER 11   THE SUPERAQUEDUCTO PROJECT – PUERTO RICO

Infrastructure Development Systems IDS-99-W-107

Research Assistants James Habyarimana and Joe Guerre prepared this case under the super-vision of Professor John B. Miller as the basis for class discussion, and not to illustrate either effective or ineffective handling of infrastructure development related issues. Data presented in the case may have been altered to simplify, focus, and to preserve individual confidentiality. The assistance of Roger Remington and Thames Water International are gratefully acknowledged.

Introduction

The Puerto Rico Aqueduct and Sewer Authority (PRASA), a public company of the Commonwealth of Puerto Rico, is responsible for the treatment and distribution of potable water and for the collection and treatment of wastewater throughout Puerto Rico. In recent years, the Island's population growth has brought an increase in water demand through enhanced domestic, commercial, industrial, and tourism activities. In the past, PRASA had been unable to counter the growing demand with increased potable water supplies. The problem culminated in 1994, when severe drought conditions resulted in inadequate water supply to approximately 2,000,000 people in the San Juan Metropolitan Area for several months. In an effort to insure sufficient water supply to the North Coast areas of Puerto Rico in the future, PRASA issued a Request for Proposal (RFP) for the North Coast Superaqueduct Project.

The Project objectives as listed in the RFP were as follows:

- Provide potable water of the required quality and quantity from the Dos Bocas Water Basin and/or the Rio Grande de Arecibo (40 miles west of San Juan) to North Coast Municipalities from Arecibo to the San Juan Metropolitan area, Caguas, Bayamon, and part of Carolina (see map below). All supplies of water were to meet or exceed the current potable water standards established by federal and local drinking water authorities, including but not limited to the Safe Drinking Water Act, the proposed Enhanced Surface Water Rule, and the Disinfection Byproduct Rule.

- Integrate the Project's water treatment plants with the Sergio Cuevas, Enrique Ortega, and Guaynabo water treatment plants in the San Juan Metropolitan Area.

- Provide the necessary interconnections to the North Coast Municipalities along the transmission pipeline.

- Provide the necessary flexibility in the metropolitan aqueduct system to allow extended maintenance of the existing facilities.

- Provide the most cost effective and reliable water supply to the North Coast Municipalities.

- Optimize the use of energy required to operate the Project's facilities.

- Optimize the operational efficiency of the Project, including the use of computers and remote control devices.

- Comply with all federal and local environmental laws and minimize and/or mitigate adverse environmental impacts.

- Create a system that has the capacity and the flexibility to provide increased amounts of potable water so as to satisfy demand through the year 2050.

- Commence Project operation as soon as practicable, and in any event on or before May 30, 1999.

Request for Proposals (RFP)

The scope of the work described in the RFP included planning, design, construction, operation, and maintenance of a regional aqueduct system to meet actual and projected potable water demands along the North Coast Municipalities. Regardless of the terms of the Master Agreement (MA), the

Project was intended to include one or more intake facilities, one regional water treatment plant, the necessary transmission lines and pumping stations, and all other auxiliary and related facilities. The intent of the RFP was to provide responding consortia with *maximum flexibility* to design a regional aqueduct system, which satisfied PRASA's needs and requirements. The selected consortium would oversee the operation of the Project for a Term of 10 years (after substantial project completion; defined as the date on which the Project meets the minimum acceptable standards set by PRASA). PRASA reserved the option to extend the Term for an additional 5 years.

Exhibit 11-1   Route of the Superaqueducto Along the North Coast of Puerto Rico

The RFP stated that PRASA would evaluate the proposals taking into account the following criteria:

- Feasibility of the consortium's recommended Project plan: This criteria included the most economic price of delivered water, the most effective capital investment for the expected life of the Project, and the minimization of the environmental impact with respect to the construction. Additionally, the longevity and reliability of the components, the ability to construct the Project on time and within budget, and the ability to operate and maintain the Project in a cost effective, safe, and reliable manner were crucial features of the feasibility evaluation.

- Capabilities and related experience of the consortium: This criteria included the evaluation of the member companies, its project team, its sub-contractors, and the individuals responsible for the project management of each phase of the Project. Long term planning, training, monitoring, operations management, and the maintenance of quality, safety, and security throughout construction, operations and maintenance phases of the Project were also evaluated. The responding consortia's ability to provide appropriate liaison with all government agencies involved with the Project was also important.

The RFP required that Puerto Rican firms, which were registered and had been doing business in PR for 10 years, perform at least 33% of the work (in each phase of the Project).

The RFP also included several quality assurance measures for this project. These measures included:

- Retainage fee: A retainage fee of 10% of the design-build price was required to ensure acceptable performance at the construction phase.

- Damages for Delays: The Company shall pay PRASA $30,000 per day in damages for delays in the completion of the construction of the Project. A maximum limit on the damages was set at $10,950,000 (just under a year of overruns).

- Bonus Payments for Early Completion: The Company was entitled to $15,000 per calendar day for every full day that the Project achieves substantial project completion prior to the substantial project completion deadline. However, the bonus was subject to an aggregate limit of $3,000,000.

- Liquidated Performance Payments: The table below illustrates the liquidated performance payment schedule that was described in the RFP.

Event	Liquidated Payment
Failure for 1, 2, or 3 days in a month	None
Failure for 4 or 5 days in a month	5% of the monthly fixed fee
Failure for 6 or 7 days in a month	10% of the monthly fixed fee
Failure for 7 or more days in month	15% of the monthly fixed fee

Table 11-1   Liquidated Performance Payment Schedule

- Performance Bonds: A performance bond in the amount equal to the max of either 50% of the design-build cost or $150,000,000 was required to guarantee the faithful performance of the construction of the Project and the debts to labor, materials, and equipment. Similarly, PRASA required a performance bond equal to the annual fixed fee for operations and maintenance.

- Warranties: The RFP described a one-year warranty on all the equipment and infrastructure after the final transfer of the facility back to PRASA.

The Winning Consortium

The criteria listed in the RFP enabled PRASA to award the contract to the consortium that offered the best value instead of basing the award solely on the lowest priced proposal. On January 31, 1996, PRASA awarded the contract to a consortium lead by Thames Water International (TWI) although another bidder had submitted a lower overall price. Two articles in the Appendix address a lawsuit that ensued between PRASA and the low bidder. Below is a list of the major project members of the winning consortium with a brief description of their roles in the Project.

TWI

TWI is based in the United Kingdom and has recently overseen the successful deregulation of the water services provision in the UK. TWI was one of the consortium co-leaders and is the leading firm in the operation and maintenance of the Project.

Dick Corporation

Based in Pittsburgh, PA Dick Corporation is primarily a construction services firm. They had total construction responsibility of the Project. Their responsibilities included entering into sub-contractual relationships with other firms. Dick Corp, in conjunction with the design engineers, was also responsible for the identification of design concepts that lead to the most optimal lifecycle cost of the Project based on technical parameters.

Quinones, Diez, Silva y Associados (QDSA)

QDSA was responsible for the design engineering of the main transmission pipeline, the pump stations, river intake structure, the raw water storage reservoir, electrical transmission lines, and related infrastructure.   QDSA was also responsible for the design of interconnections (to towns along the pipeline route) and the treated water reservoirs at these interconnections.

Vincenty, Heres y Lauria (VHL)

VHL is the second of the design-engineering firms.   VHL was responsible for the design of the water treatment plant and related infrastructure.

The project team also included a number of project advisors.   Most notably, Chase Investment Bank, the financial advisor, and PWT Projects (a subsidiary of TWI), the process advisor to the consortium.   In an attempt by the consortium to meet the proposal criteria, the principal design-engineering firms are based in PR.   However, TWI had retained the services of Chiang, Patel and Associates as a design advisor.   PRASA also reserved the right to appoint an engineering consultant as its representative during the design, construction, and operation of the Project.   Figure 11-1 below illustrates the organizational structure for the Project.   The arrows indicate contractual relationships.

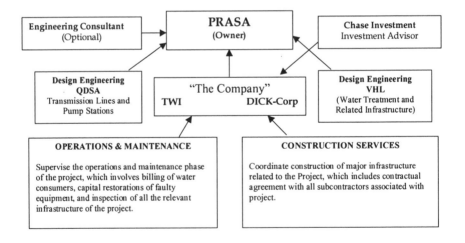

Figure 11-1   Organizational Structure of Project

The MA clearly stated that PRASA would retain the company and engineers on an independent basis. The consortium was therefore not a joint venture, but rather a number of firms each with a specific function within the Project. The Company, however, was the chief guarantor of the engineering firms. The MA also stated that PRASA is to have no contractual relationship with any sub-contractors. The Company was therefore responsible for all the risk of default or non-satisfactory performance of the engineers and subcontractors.

The Winning Proposal

The elements of the project as described in the winning proposal are summarized below.

Raw Water Reservoir: The water system will enter the system via a 61m long reinforced concrete weir located along the Rio Grande de Arecibo (In Figure 11-1, this river is shown but not labeled. Rio Grande de Arecibo empties into the Atlantic Ocean near the city of Arecibo.) The river will spill over the weir at a rate of 104 million gallons (MG) per hour at 1m head into a 300-MG storage reservoir. The reservoir will be strategically located within a natural oxbow of the river. The location of the reservoir enabled the engineers to locate a second weir with two sluice gates on the downstream side of the reservoir. The second weir has two functions. First, it will act as the outlet for the reservoir. Secondly, the sluice gates will act as second inlet structure during periods of low river flow.

Raw Water Pump Station: Water from the reservoir will spill over another intake weir structure into a pump station located at the northeast corner of the reservoir. The pump station will have a capacity of 100 MGD. The station is designed to house 5 vertical turbine pumps, each with a capacity of 27 MGD at 117 m. The pump station design also includes separate buildings to house the motor control center, 4 backup generators each with a capacity of 1,800Kw, and electrical switchgears. The generators are to be fed by 2 fuel storage tanks. This design permits the station to operate during electrical failures.

Raw Water Pipeline: The pump station will send water through 7 km of 72-inch reinforced concrete pipe. The force main has a flow capacity of 100 MGD at 5.5 fps. The lines will be installed with a minimum cover of 4 feet, and air-release valves will be located at all high points and blow-off valves will be located at all low points.

Water Treatment Plant: The raw water main will lead to a 100 MGD water treatment plant. The plant design includes 2 coagulation mixing tanks, 8 upflow clarifiers, 16 dual media filters, 2 chlorine contact tanks, 4 washwater recovery tanks, 2 sludge lagoons, and a treated water storage tank.

The plant will be built in a sparsely populated area in order to minimize the residential impacts of the facility. The plant's elevation will permit treated water to be distributed by gravity rather than by additional pumping facilities.

Treated Water Pipelines: The potable water from the treatment plant will be discharged into two gravity pipelines, the Arecibo Reach and the Bayamon Reach. Approximately 70% of the lines will be installed along existing highway right-of-way. The proposed location will be easily accessible during construction and maintenance and minimize ROW requirements and environmental impacts.

The Arecibo Reach will flow northwest back to the city of Arecibo through a 36-inch RCP. This reach has a capacity of 15 MGD. The Bayamon Reach will flow east towards the City of San Juan through a 66-inch RCP for a total of 57 km. This reach has a capacity of 75 MGD. The layout of the gravity line follows the path of a major state highway when possible, in order to minimize the environmental and residential impact of the project. The Bayamon Reach will be connected to a series of two inline prestressed concrete storage tanks, each with a capacity of 10 MG. The main will also be interconnected to the following municipalities:

Interconnection Flow Capacity (MGD)	
Barceloneta	3.6
Manati	7.3
Vega Baja	9.4
Vega Alta	6.7
Dorado/Toa Baja	18.1
Toa Alta	9.4
Bayamon South Reach	75

Table 11-2   *Interconnection Flow Capacity*

Work Force: The work force that will construct the project will consist of 80% local forces. Local workers will also account for 90% of the labor during the operations phase.

Project Schedule and Budget

Design/Build Schedule

The substantial project completion deadline date (SPCDD) of the Project was defined as the latest of either the design notice to proceed (DNP) + 29 months, the procurement notice to proceed + 27 months, or the construction notice to proceed + 25 months. The final project completion deadline date (after which operation begins) was defined as SPCDD + 60 days. The DNP was issued in February 1996.

Cost: Design-Build

The cost of design and construction was determined using the "cost-plus" approach that yielded a Guaranteed Maximum Price (GMP) of $264,740,000.00. The GMP included the profit of the parties involved as well as contingency. The GMP was fixed at a predetermined figure established during the competition phase. The Design engineering firms were paid lump sum design service fees. These fees could only have been changed if a change in the scope of work was mutually agreed to between PRASA, the Engineers, and the Company. A break down of the GMP is shown below.

A.	Cost of Construction Services	$181,113,800.00
B.	Construction Services Fee	$36,375,000.00
C.	Contingency	$31,038,000.00

Design Services Fee

1.	VHL Design Services fee	$7,532,180.00
2	QDSA Design services fee	$8,681,020.00
	Guaranteed Maximum Price (GMP)	$264,740,000.00

Contingency funds could only have been used to meet costs associated with the acceleration of the construction phase or the cost involved in an increased scope of the Project mutually agreed upon by the PRASA and the Company.

Cost: Operate-Maintain

The Operation and Maintenance Services fee covered all costs required to operate and maintain the completed facility. The Operation and Maintenance fee consisted of the following:

A fixed fee covered the cost of all operation and maintenance services excluding pass through costs. Sludge disposal and plant maintenance were defined as fixed costs.

Pass through costs were variable costs including payments for power, chemicals, and fuel consumption.

The Operations and Maintenance fee was tied to the annual consumer price index calculated by the Puerto Rico Department of Labor.

The breakdown of Operations and Maintenance fee is as follows:

Item	Price per year	Price per month
A. Fixed Fee	$4,484,244	$ 373,687
B. Estimated Pass through costs	$5,853,948	$487,829
Total	**$10, 338, 192**	**$861, 516**

Project Financing

The RFP stated that the project could be financed publicly, but also provided the bidders with the option to submit proposals using private funding. In the end, the project was financed by PRASA. Revenues were raised by the Commonwealth of Puerto Rico and combined with Federal funding. In the following questions, assume that 25% of the GMP was provided directly to PRASA through a federal grant at the start of construction. This amount need not be repaid. Assume further that the Commonwealth of Puerto Rico had the capability of raising the balance of funds required for design and construction through the sale of 20-year tax-exempt bonds sold in the market at an annual borrowing rate of 6.0%.

Upon completion of the project, revenues will be collected from payment for the water produced. There was no mention of the price of water to be charged by the Company in the agreement. However, PRASA can determine the price using two basic scenarios.

• Scenario 1: PRASA chooses not to extend the 10-year Term.

- Scenario 2: PRASA chooses to exercise their option to extend the Term by 5 years.

The Consortium would be paid on a "percentage completed" basis throughout the design and construction period. Upon completion, O&M will be paid for with the monthly payment described above.

Two ENR articles from February, 1996 provide additional background.

Exhibit 11-2 ENR Editorial, "If You Don't Shoot Straight, You May Hit Your Own Foot"
Exhibit 11-3 ENR Article, "Contract Award Challenged"

Both Exhibit 11-2 and 11-3 are Reprinted from Engineering News-Record, copyright The McGraw-Hill Companies, Inc., February 26, 1996, All Rights Reserved.

Questions

1. Where does the project fit in the Quadrant framework? Why?

2. Review the information about the Project's financing in the case. Pick twenty (20) years from the end of construction as a common period for analysis and prepare a cash flow from the perspective of PRASA for two scenarios: 1) PRASA does not extend the original 10-year term, and 2) PRASA elects to extend the first term by 5 years. For each scenario, use the same time period and determine the minimum price of water necessary to make the project viable by the end of the 20-year operation term. What assumptions did you have to make to complete each cash flow?

   Use the following design and construction schedule shown below (% of GMP) and assume that full operation starts in the 3rd quarter of the third year.

Year	1	2	3
% GMP	28.5	59	12.5

Assume further that the initial water demand in the 3rd quarter of the third year is projected to be 54.3 million gallons per day (MGD), and is estimated to increase quarterly by 0.36%. However, remember that the capacity of the system is 90 MGD.

Assume that the initial annual O&M cost is $10,338,192 and that it increases annually by a rate of 3%. This increase reflects the expected changes in the pass-through costs. Assume further that PRASA can operate and maintain the system at 107% of the cost of the consortium.

   a.   What discount rate did you use in your analyses? Why?

   b.   Evaluate the contractual arrangement. What advantages and disadvantages do you see?

3.   Assume here that instead a 20-year franchise from the end of construction is offered to private sector competitors in which the successful bidder must raise its own funds for design and construction of the facility in the capital markets. Assume that the federal grant money is available and unchanged in timing or amount (i.e 25% of the GMP cost for design and construction). In connection with this scenario, assume that there is no further reduction in the initial cost of the facility and that tax-exempt status is available under applicable IRS rules that permit private management and operation (but not ownership) of public facilities. Prepare a second cash flow analysis of the Project, again from the perspective of PRASA. Assume that the franchise borrowing rate is 8% annually.

   a.   If the price of water is consistent with the price calculated previously for Question 2, Scenario 1, and the company's annual weighted average cost of capital is 15%, i.e. the discount rate that they use for project evaluation, what must the NPV of the company's actual design-build costs be for the company to conclude that this Project is viable?

    b. Is private financing feasible for this project?   Explain your
        answer.

References

"If you don't shoot straight, you may hit you own foot." Engineering News
    Record Feb 26, 1996: 70.

Master Agreement for the North Coast Superaqueduct Project by and among
    PRASA, Thames-Dick Superaqueduct Partners, Inc., Vicenty, Here Y
    Lauria, Sociedad de ingenieros ambientales, and Quinones, Diez, Silva
    y Asociados, Jan. 31, 1996.

Buchanan, Barry. North Coast Superaqueduct Project: PUERTO RICO,
    Technical features of the Project. (Presentation notes.)

North Coast Superaqueduct Project Request for Proposals, with
    amendments. Issued by PRASA. Feb. 17, 1995.

Puerto Rico Map. Mar. 15, 1995
        www.lib.utexas.edu Libs PC Map_collection/americas/PuertoRic
    o.jpg>.

Thames-Dick Superaqueduct Partners Inc. Technical Proposal. Jan. 17,
    1996: Appendix 2.

CHAPTER 12   MANAGED PUBLIC/PRIVATE COMPETITION FOR WASTE WATER TREATMENT WORKS

Charlotte-Mecklenburg Vest Water Treatment and Irwin Creek Wastewater Treatment Plants

Infrastructure Development Systems IDS-97-W-103

Research Assistant Maia A. Hansen prepared this case under the supervision of Professor John B. Miller as the basis for class discussion, and not to illustrate either effective or ineffective handling of infrastructure development related issues. Data presented in the case has been altered to simplify, focus, and to preserve individual confidentiality. The assistance of Kimberly Eagle and Barry Gullet of the CMUD, Ann Fernicola and David McKinsey of U.S. Water, Mike Gagliardo of the U.S. Conference of Mayors, Robert G. Joseph of U.S. Filter Corp., and David Zimmer of Camp Dresser & McKee in the preparation of this case is gratefully acknowledged.

Two Perspectives on the Charlotte Competition:

"The three CMUD projects were examples of progressive leadership and the potential that exists in the introduction of competition in a public works environment. It should be understood however, that CMUD did not simply submit its best guess at a competitive bid. CMUD used the best techniques of both the public and private world to eliminate inefficiency and to add financial incentives to a public employment setting. The results were dramatic with more than $2.2 million in savings beyond those that would have been realized had CMUD not participated in the process. Thanks to a well run process where the public employees were equipped to do their best, the BIG winners in Charlotte and Mecklenburg County were the rate payers that will benefit from the savings."

-Trille C. Mendenhall, Charlotte Mecklenburg Utility Department (CMUD) and John F. Williams, HDR Engineering, Inc.[28]

"The Charlotte-Mecklenburg Utility Department submitted the lowest bid to operate the Vest Water Treatment Plant and Irwin Creek Wastewater Treatment Plant.

CMUD's bid of $7.1 million was lower than [a] an $8.8 million bid submitted jointly by Denver-based JMM Operational Services and J.A. Jones Management Services Inc. of Charlotte; [b] Englewood, Colo.- based OMI Inc.'s $9.3 million bid, and [c] Duke Engineering & Services Inc.'s $9.7 million. Duke joined with Charlotte Water Services and American Anglian Environmental Technologies in its bid.

Other bidders included: Wheelabrator EOS Inc. of Hampton, N.H., $10.9 million; U.S. Water, $12 million; and Houston-based PSG- Professional Service Group, $14.6 million.

J.A. Jones says it was surprised to find the city with a lower bid because it submitted a bid that was 20% lower than what the CMUD now spends to operate the plants.

'And now the city number is about 16% below that (bid)," says J.A. Jones spokeswoman Ede Graves. "We think perhaps there were some items left out of the city's bid inadvertently or whatever. The contracts were supposed to include operating and maintenance and we're thinking they didn't put the maintenance cost in.'"

Brian Gott, *Lowest bid is CMUD's*, The Business Journal of Charlotte, May 6, 1996, SECTION: Vol. 11; No 4; pg. 8

History of Charlotte

Charlotte, North Carolina is the seat of Mecklenburg County and the largest metropolitan area in the Carolinas. As of 1990, Charlotte became the

35th largest city in the U.S., showing unusual growth compared to other urban areas. The percentage increase in population was 30.2% in the 1970s and 25.5% in the 1980s. In 1995, the Charlotte population had grown to 579,473 and it accounted for most of the county population and area.

With a well-diversified range of industries, including one of the nation's largest textile centers, railroad and distributing centers, extensive manufacturing, food processing, and banking, Charlotte is a rapidly growing metropolitan area. The largest manufacturers are Frito-Lay, Coca-Cola, Celanese, General Tire and Rubber, and IBM. In addition, as an inland port facility and a foreign trade zone, Charlotte attracts foreign investors looking for a foothold in the Southeast. By 1991, 260 foreign-owned companies had facilities in Charlotte.

Charlotte is led by a council-manager form of government. The annual city budget (general fund), in 1995, consisted of revenues of $216 million and expenditures of $196.5 million. The debt outstanding was $666 million in general obligation bonds. Total taxes per Capita were $292. This was broken down into $240 for property taxes and $52 for Sales and Gross Receipts.[29] Exhibit 12-1 shows general statistics about the economy of Charlotte in the early 1990's. The city maintains a triple "A" rating by Moody's Municipal & Government Manual.

Exhibit 12-1   Summary of Charlotte-Mecklenburg Economic Statistics

Charlotte Mecklenburg Utility District

In 1972, the City of Charlotte and the County of Mecklenburg restructured their administrations and combined their water treatment, water distribution, wastewater collection, and wastewater treatment services under one title, the Charlotte Mecklenburg Utility District (CMUD). As the area developed CMUD expanded its role to include water and wastewater treatment systems for many small towns in Mecklenburg County and beyond to Union County. As of 1995, Douglas Bean was the head of the department, with Barry Gullet as deputy director. An organizational chart is shown as Figure 12-1. The CMUD owns and operates three water treatment plants and five wastewater treatment plants, which range in age from newly constructed to 75 years old. A summary table of plants is shown below in Table 12-1. The water system has a total plant processing capacity of 138.8 million gallons per day mgd and the wastewater systems capacity totals 92 mgd, which serves 550,000 residents and industries of the city and county as well as many small surrounding municipalities.

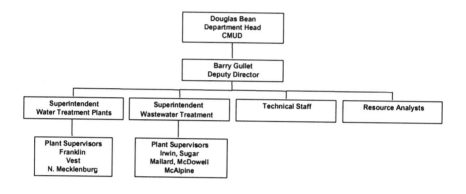

Figure 12-1　　*CMUD Organizational Structure During Normal Operations*

Plant Name	Year Completed	Capacity (mgd)
Irwin Creek Wastewater Treatment WWTP	1927 (1948, 1973)	15
Mallard Creek WWTP	1980	6
McAlpine WWTP	1966 (1974, 1981, 1982, 1994, 1996)	48
McDowell Creek WWTP	1979 (1994)	3
Sugar Creek WWTP	1927 (1951, 1960, 1995)	20
Franklin Water Treatment Plant	1959 (1967, 1981, 1987, 1990)	96
Vest Water Treatment Plant	1924 (1939, 1948)	24.8
North Mecklenburg Water Treatment Plant	1996	18

Table 12-1　CMUD's Water and Wastewater Treatment Plants

Wheelabrator Purchase Proposal

In January 1995, as part of a corporate strategy to expand its position in privately operated U.S. water treatment systems, Wheelabrator EOS proposed to buy the large McAlpine Creek wastewater treatment plant from the city of Charlotte. Under the authority of Executive Order 12803 (See Exhibit 12-2 on the companion CD), the sale of municipal infrastructure to a private company had become simpler and less costly to cities. Wheelabrator was in the midst of completing the Franklin, Ohio wastewater treatment plant deal, the first wastewater plant sale in the country, and it wanted to continue its success to other areas of the country.

Exhibit 12-2   President George Bush, Executive Order 12803

Charlotte considered this unsolicited proposal and decided, after much discussion, not to accept it. It conducted a survey of other cities' decisions to sell or lease infrastructure, and concluded that there were three general reasons for privatization: 1) the inability to raise capital easily, particularly in small cities; 2) financial crises in large cities; and 3) labor union problems. The leaders in Charlotte felt none of these needs were applicable to its circumstances. The city was solvent, active in the capital markets, and on good terms with its employees.

In addition to the lack of a catalyst for privatization, the city found numerous legal and administrative restrictions to the sale of its wastewater treatment plant. These included: 1) EPA and state concerns about the transfer of National Pollution Discharge Elimination System (NPDES) permits to a private organization; 2) potential withdrawal of current EPA funding for upgrades at one of the city's larger plants, if federal loans were expected to be forgiven under Executive Order 12803, 3) regulation of the plant as an industry under RCRA (higher standards) rather than as a municipality under the Clean Water Act; 4) uncertainty about immediate payment due to the EPA or state for construction grant money; 5) the loss of the plant's tax exempt status on current and future bonds; and 6) the question of whether a public vote was required by the state before infrastructure could be sold.

Furthermore, analysts for the city saw the sale of the plant as little more than an option to increase cash, with a premium paid through the cost of capital of the private firm. As described in a briefing to the city: "The most crucial point that needs to (be) made in order to understand this concept (of selling the plant) is the understanding that the proceeds from the sale of the plant which are going to the City are in reality a borrowing since the private company's cost of capital will be included in the service contract. Unlike the sale of a coliseum or other city property where the city is getting out of business, the private owner of a wastewater treatment facility will still have the city as its sole customer. Any business has to recover its capital costs, therefore, the purchase price will be included in the service fee which the city will be paying to the private operator" - Preliminary Feasibility Study of a Sale of the McAlpine Creek Wastewater Treatment Plant, CMUD.

Despite rejecting the Wheelabrator offer, the city administration remained interested in the concept of competition for operations and maintenance of the water and wastewater systems. The much-publicized competition for services in cities like Indianapolis, Indiana had shown significant cost savings through increased efficiency. Since February 1994, Charlotte's leaders investigated a number of possibilities to invite more competition and effect cost savings in the operation of municipal services. Most of the services had been small, albeit successful, projects, such as garbage collection, landscaping, and printing and copying. The council appointed a taskforce, called the "Competition / Privatization Advisory Committee," to determine the potential for competition in the city. Through this committee, the city developed a comprehensive 5-year program to compete a broad range of services. (See Table 12-2)

Table 12-2 Notes: All costs are annualized. Services are listed in the year in which the RFP will be issued. Contract amounts are actual for FYs 1995 and 1996, estimated for FYs 1997, 1998, and 1999. Savings are estimated.
*Maintenance of instrumentation at one plant scheduled to be contracted out as a benchmarking effort.

Services Available in Private Sector	Employees at Risk If CMUD Loses Bid	Contract Amount ($)	Savings Estimated ($)
Scheduled for Competitive Bidding in FY 1995			
Odor control	1.0	125,000	225,000
Maintenance of grounds	3.0	99,590	39,353
Janitorial services	3.0	66,708	34,318
Total	**7.0**	**291,000**	**298,000**
Scheduled for Competitive Bidding in FY 1996			
Irwin Creek Wastewater Treatment Plant	25.0	2,700,000	769,858
Use of wastewater byproducts as fertilizer	0.0	680,000	164,300
Residuals Compost Management Facility	6.0	1,300,000	186,000
Vest Waster Treatment Plant	14.0	1,200,000	193,993
Location of water lines	3.0	160,000	0
Testing of soils and materials	5.0	192,750	1,000
Maintenance of right-of-way	5.0	100,100	19,040
Maintenance of lift stations	9.0	665,000	0
Maintenance of instrumentation*	2.0	111,600	N/A
Total	**69.0**	**7,109,450**	**1,361,191**
Scheduled for Competitive Bidding in FY 1997			
Labs	12.0	775,000	
Hydraulic and mechanical sewer cleaning	20.0	733,000	
Service renewals	5.0	450,000	
Trunk line monitoring/basin	0.0	250,000	
Permanent flow monitoring	0.5	65,000	No
Meter reading (1/4 route sets)	6.0	198,000	Info.
Herbicide/TV sewer lines (cleaning)	3.0	140,000	Available
TVing sewer lines (cleaning)	3.0	121,000	
Total	**49.5**	**2,732,000**	
Scheduled for Competitive Bidding in FY 1998			
Maintenance of hydrants	8.0	478,000	
Maintenance of meters (1/4 system)	2.0	63,000	
Total	**10.0**	**541,000**	
Scheduled for Competitive Bidding in FY 1999			
Preventive and corrective maintenance	19.0	1,260,000	
Water main repairs	5.0	128,000	
Meter reading (1/4 route sets)	6.0	198,000	
Total	30.0	1,586,000	
Grand Total	**165.5**	**12,259,748**	**1,659,862**

Table 12-2   *Charlotte-Mecklenburg Five-Year Competition*
Source: *Popular Government, Winter 1997, p. 17*

The advisory committee defined its main goals of competition as follows:

- To determine the most cost effective contractor

- To establish a level playing field between private and public proposers

- To allow for meaningful proposer input

- To develop and utilize objective evaluation criteria

- To involve Citizen Advisory Boards as a resource

Exhibit 12-3 presents Charlotte's stated policy relative to competitive bidding.

Exhibit 12-3   Charlotte's Policy and Goals for Competitive Bidding

Development of the Bidding Process

The city identified three CMUD facilities that it would open for competition for operations and maintenance. The facilities were the Vest Water Treatment Plant, the Irwin Creek Wastewater Treatment Plant, and the residuals management facility. The competition was announced publicly. Excerpts from the RFP are provided as Exhibit 12-4. Private companies as well as the existing public management were invited to compete. Because of concerns for an objective evaluation, the staff of the CMUD was divided into two groups - procurement and bidding. (See Figure 12-2 for a diagram) As described by Barry Gullet and Doug Bean: "City staff and elected officials strongly desired that the competition process take place on a 'level playing field'. Consequently [we] put up an imaginary wall between the team that would assemble the bidding documents and evaluate the bids and the team preparing the staff's bid."[30]

Exhibit 12-4   Excerpts from RFP

The advisory committee's competition goals acknowledged that fair evaluation criteria were crucial to the success of this competition as well as future managed competitions.

Procurement Team

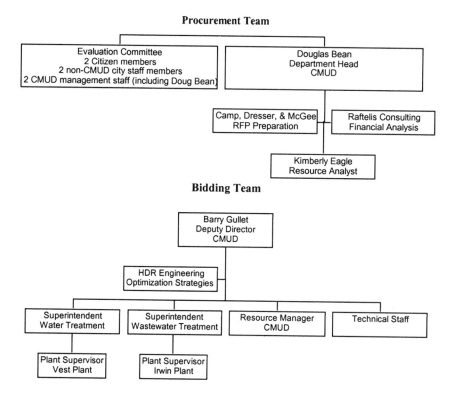

Figure 12-2   *The Invisible Wall: Organizational Charts of CMUD during Competition*

After this imaginary wall was established and the teams were in place, different outside consultants were hired to support their respective team in the competitive bid process. In order to prepare itself for the upcoming competition, the bidding team developed a ten-month optimization process to bring itself into line with operating standards at privately operated plants. Preparations included visits to other plants, training of personnel, changes in incentive and compensation systems, and cost benefit analysis in four main areas: energy and utilities, treatment processes, controls and automation, and personnel.

For both the water and wastewater treatment plants, a three-step process was used to solicit qualifications and proposals for the three-to-five year operation and maintenance contracts. Submissions for the RFQ were due on September 15, 1995 with final proposals due on April 11, 1996. The residuals management facility received little attention from the private sector, but both the Irwin Creek and Vest plants were highly sought after. Nine private firms responded to the RFQ for the Irwin plant and ten for the Vest plant. Draft RFPs were issued shortly after receipt of the RFQs, and all competitors were given time to comment, tour plants, and ask questions about the bidding process.

The Request for Proposals

Difficulties in comparing public and private bids were recognized from the start by the evaluation team. To promote even evaluations of the competing groups, the procurement team developed a detailed request for proposals (RFP) for each of the three facilities mentioned above. Bidders were asked to focus specifically on operations and maintenance activities, and were not to include capital improvement projects requiring amortization longer than the contract term. In fact, each bidder was required to use a plug value of $50,000 for corrective and heavy maintenance costs each year. Once the bidders' line item costs were received, the evaluation team discounted the costs at 5.75% per year and added in set additional costs for the bids.

These cost allocations differed between the public and private bids; for example, CMUD's bid was charged for city overhead and administration costs, while bids from private firms were told to account for pre- and post-operations conditions studies, contract negotiations and administration in their bids. CMUD was not required to provide additional insurance or the cost of performance bonds for its bid.

The Competition/Privatization committee debated CMUD's overhead charges at great length. The extent to which city and department overhead should be assigned to the bids was difficult to determine, since costs were not easily separated. Discussions continued until finally a range of overhead charges, between $32,149 and $77,362 per year, was agreed upon.

Proposer	Year	Proposed Annual Fee - Irwin	Proposed Annual Fee – Vest
	1	$1,051,227	$496,194
CMUD	2	$1,084,459	$509,694
	3	$1,106,773	$515,215
	4	$1,131,010	$527,858
	5	$1,195,847	$548,999
	Total	**$5,569,316**	**$2,597,960**
	1	$1,343,068	$634,490
OMI Incorporated	2	$1,375,354	$649,800
	3	$1,408,541	$665,562
	4	$1,442,659	$681,789
	5	$1,477,739	$698,497
	Total	**$7,047,361**	**$3,330,138**
	1	$1,645,000	$1,147,000
US Water	2	$1,655,000	$1,156,000
	3	$1,688,000	$1,180,000
Total Bid: $	4	$1,719,000	$1,202,000
	5	$1,751,000	$1,226,000
	Total	**$8,458,000**	**$5,911,000**
	1	$1,527,375	$877,905
Wheelabrator EOS Inc	2	$1,553,091	$892,103
	3	$1,579,576	$906,728
	4	$1,606,857	$921,790
	5	$1,634,957	$937,307
	Total	**$7,901,856**	**$4,535,833**
	1	$1,322,883	$689,485
JMM / Jones, Joint Venture	2	$1,347,175	$705,190
	3	$1,372,159	$721,345
	4	$1,367,793	$737,962
	5	$1,393,650	$755,054
	Total	**$6,803,660**	**$3,609,036**
	1	$1,281,165	$794,535
Duke Engineering	2	$1,310,540	$825,165
	3	$1,341,227	$857,352
	4	$1,373,292	$891,178
	5	$1,406,790	$926,731
	Total	**$6,713,014**	**$4,294,961**
	1	$3,086,096	
Professional Services Group	2	$3,167,053	
	3	$3,250,438	
(submitted one combined bid for	4	$3,336,326	
both Vest and Irwin Plants)	5	$3,424,789	
	Total	$16,264,702	
	1		$901,539
Consumer Applied Technologies	2	No bid submitted	$784,711
	3		$783,834
	4		$804,957
	5		$825,386
	Total		**$4,100,427**

Table 12-3   Final Bids

Evaluation of the Proposals

In April 1996, the CMUD procurement team received eight bids, seven from private firms and one from the CMUD bid team. Seven of the bids were for combined operation of both the water and wastewater plant. One bid was for the water plant only. A listing of the total bid amounts is shown as Table 12-3 and a breakdown of CM-ConOp's bid is detailed in Table 12-4. The bids of the losing private firms were generally considered proprietary information and not released publicly by CMUD.

COST COMPONENT	YEAR1	YEAR2	YEAR3	YEAR4	YEAR5
Wages and Salaries	$237,926	$247,443	$257,341	$267,634	$278,340
Other Personnel Costs/Benefits	$63,432	$66,906	$70,487	$74,161	$77,912
Chemicals	$55,003	$60,003	$60,003	$60,003	$65,003
Maintenance Allowance	$50,000	$50,000	$50,000	$50,000	$50,000
Electricity	$300,025	$327,300	$327,300	$327,300	$354,575
Other Utilities	$6,600	$7,488	$6,706	$6,974	$7,858
Outside Services (see attached sheets)	$202,944	$188,006	$195,526	$203,347	$218,303
Other Costs (see attached sheets)	$93,499	$95,515	$97,612	$99,792	$102,059
Overhead/Fee*	$41,798	$41,798	$41,798	$41,798	$41,798
Annual Fee	$1,051,227	$1,084,459	$1,106,773	$1,131,010	$1,195,847
Electricity Usage and Demand					
Electricity Usage (Kwh)	8,530,000	8,530,000	8,530,000	8,530,000	8,530,000
Electricity Demand (Kw) On-Peak	1,133	1,133	1,133	1,133	1,133
Chemical Usage (pounds)					
Chlorine	54,311	54,311	54,311	54,311	54,311
Sodium Bisulfite	29,147	29,147	29,147	29,147	29,147
Caustic	22,580	22,580	22,580	22,580	22,580
Polymer	69,093	69,093	69,093	69,093	69,093
*Represents Variable Portion of Department and Division Overhead.					

Table 12-4   CMUD's Bid by Section

The six person evaluation team, consisting of one citizen member of the City of Charlotte Competition / Privatization Advisory Committee, one citizen member of the CMUD Advisory Committee, two non-CMUD members of city staff, and two members of CMUD's management staff, found that CM-ConOp proposed both the lowest cost as well as the best technical approach to operations and maintenance. Although the employees were not required to sign a contract following the competition, they volunteered to sign a memorandum of understanding (MOU) with the city.

A copy of the MOU for the Irwin plant is included as Exhibit 12-5. The contract went into effect July 1, 1996, after being approved by the City Council.

Exhibit 12-5   Memorandum of Understanding between the City and CMUD Employees

Trill Mendenhall, of CMUD, and John Williams, of HDR Engineering, described the results this way: "CMUD's bid received the highest technical rating and was also the lowest bid over the five-year contract term. ... The proposals received included staffing, chemical consumption and residuals management elements that were remarkably similar. ... [The] spread between CMUD's offering and the next bidder matches closely with the additional costs that are inherent to the private sector (higher overhead, taxes, administrative charges, travel, insurance, performance bonds, and rampup costs)."[31]

In summary, the city of Charlotte chose to open-up water and wastewater privatization to local city management as well as large private contract operations firms. The result was a competition that drew intense participation on all sides. The citizens were rewarded with noteworthy cost savings, more efficient plants, and internal benchmarking capabilities for future competitions on municipal services.

The private sector, however, is still not certain how to approach similar competitions in the future. As reported in Engineering News Record's Sept. 23, 1996 issue: "Some privatization executives indicate that they will avoid cities with managed competition, expressing fears that the "playing field" for such procurements may not be level and that in-house bidders may not have to provide the same guarantees as outsiders. 'If we see a proposal that asks for both privatization and a traditional bid, this signals no real commitment to privatization,' says Christy Cooper, director of project finance for Black & Veatch, Kansas City."

Reflecting on the competition a year after the award, Ann Fernicola, the Project Manager for the U.S. Water bid team, stated: "This was the first time we had seen something like that - where the city was one of the competitors. We were never completely comfortable with how it worked and if the level playing field really was there. The city had a couple of meetings where they attempted to explain what was and was not included in the public and private bids. They talked about the differing treatments in labor, for example, where they were hindered by existing labor contracts. The financial details were never discussed, though. Specifically, there was no discussion before or after the competition on how the maintenance and administrative sections were counted. For example, the city had the opportunity to utilize existing Public Works services for maintenance, whereas a private contractor didn't." Although she felt Charlotte's bid and procurement teams did maintain the promised separation of information, she thought the process could have been more open. "Charlotte sent us a second invitation to bid on their latest project. We've already sent them a curt 'no'."

David McKinsey, also of U.S. Water, felt that a truly level playing field could not be created in this type of procurement. He said, "You have to either 1) make the public side adhere to all requirements or 2) allow the private sector the same latitude allowed the public. The public sector is not ready to compete in the first option and the second option takes away many of the financial and risk management tools intended by privatization in the first place." Although he was not personally involved in the bidding, he "sensed that the requirements were pretty well explained. Everyone knew that the public side was bidding and they still bid. I guess we didn't think the public guys would be as smart as they were in reducing costs." The cost of bidding on a project like this was estimated at $100,000 to $150,000 per competition. In summary, he said he would not bid on a similarly structured project in the future.

Closing

"The Charlotte model for competition showed that the public sector can compete successfully with private firms and that the result can be a complete change in the way the agency does business...Often, the truth is that under existing circumstances where public departments may be forced to adhere to outdated work rules, payment policies and bloated workforces, private companies can offer the potential for savings. However, when allowed to function like a private company in a competitive environment,

there is no reason why a municipal utility cannot function as efficiently. To be successful, the public employees must be allowed to apply innovative "private" approaches such as cross-training employees, financial incentive, and performance management.   Also, the utility system needs to be financially and technically stable."

-Trille C. Mendenhall, CMUD and John F. Williams, HDR Engineering. Inc., Paper for the Water Environment Federation convention.

"The city of Charlotte recently issued a "Request for Proposal" for operation and maintenance services of its Irwin Creek wastewater and Vest water treatment plants. Unlike other municipalities, however, the city's own utility department was allowed to submit a competitive bid to continue operations.

*After careful review of the bids, Charlotte's City Council voted to accept Charlotte-Mecklenburg Utility Department's (**CMUD**) proposal, and in doing so, set a national precedent. The winning bid submitted by its own **CMUD** was 20 percent lower than the closest private firm seeking the contract. It proposed to reduce operating and maintenance costs by more than $4 million over the five-year contract period.*

CMUD proposed to reduce the operations and maintenance costs at both facilities by nearly $1 million by reducing personnel and enhancing automation. Cost and performance guarantees were also included in the city's proposal-language usually seen only in private service agreements."[32]

Questions

1. Evaluate the "three general reasons for privatization" offered by the City. Are they comprehensive descriptions of the interests of the City of its employees of its water/sewer users?

2. Evaluate the first five of the City's six "restrictions" identified as grounds not to accept WEOS's January, 1995 Purchase Proposal. Compare these with the actual mechanisms followed in the Franklin

and Indianapolis cases. Which "restrictions" are real, if any? How could they be avoided?

3. Where in the City's procurement process did it obtain head to head competition on price? on quality?

4. Review Exhibit 12-5, the MOU signed by the city, and compare/contrast the contractual commitments sought from private bidders with those signed by the winner.

5. What are the risks and benefits in allowing the private and public sectors to compete together for a municipal contract?

6. What are the challenges of creating a level playing field? Was a level playing field created here? What was compared on that field?

7. Evaluate the City's "10 month optimization process". What happened to the "playing field" during this period?

8. Evaluate the City's decision to separately compete the three facilities.

9. In what ways would you have followed or changed Charlotte's strategies in structuring the competition?

10. What advice would you give to Charlotte as the city prepares for the upcoming competitions on the five-year plan?

References

Mendenhall, Trille C., and John F. Williams. "Benefiting From Private Sector Techniques." Paper presented to the Water Environment Federation Annual Conference & Exposition, October 1996.

Kurian, George Thomas. World Encyclopedia of Cities. ABC-CLIO, 1994: 179.

Gullet, Barry and Doug Bean. "Charlotte Model for Competition: A Case Study." Popular Government 62.2, Winter 1997: 19.

Mendenhall, Trille C. and John F. Williams. "Charlotte Mecklenburg Utility Department (CMUD)." Paper presented to the Water Environment Federation Annual Conference & Exposition, October 1996.

Anderson, Jeanne, "Charlotte-Mecklenberg utility successfully competes against private sector." Water Engineering & Management. 143.11, November 1996: 8.

Notes

[28] Trille C. Mendenhall, John F. Williams, "Benefiting From Private Sector Techniques", Paper presented to the Water Environment Federation Annual Conference & Exposition, October 1996.

[29] George Thomas Kurian, "World Encyclopedia of Cities", ABC-CLIO, 1994, p. 179.

[30] Barry Gullet and Doug Bean, "Charlotte Model for Competition: A Case Study," *Popular Government,* Winter 1997, Vol. 62, No. 2, p. 19.

[31] Trille C. Mendenhall, Charlotte Mecklenburg Utility Department (CMUD) and John F. Williams, Paper presented to the Water Environment Federation Annual Conference & Exposition, October 1996.

[32] Anderson, Jeanne, November 1996, "Charlotte-Mecklenberg utility successfully competes against private sector", *Water Engineering & Management*, SECTION: Vol. 143, No. 11 Pg. 8.

CHAPTER 13 THE STRATEGIC GOALS OF PUBLIC INFRASTRUCTURE PROCUREMENT: WILMINGTON DELAWARE WASTEWATER TREATMENT PROCUREMENT

Infrastructure Development Systems IDS-97-W-104

This case was prepared by Professor John B. Miller with the help of Research Assistant Maia A. Hansen as the basis for class discussion, and not to illustrate either effective or ineffective handling of infrastructure development related issues. Data presented in the case has been intentionally altered to simplify, focus, and to preserve individual confidentiality. The assistance of G. Robert Joseph of U.S. Filter Corp., Tom Noyes, Assistant to the Mayor, Wilmington, DE, and Mike Gagliardo, U.S. Conference of Mayors in the preparation of this case is gratefully acknowledged.

"We are breaking new ground. We don't have anywhere else to turn to structure a contract for what we want to do."

- Tom Noyes, Executive Assistant to the Mayor of Wilmington, September 1996 Engineering News Record.

Introduction

In 1995, the nation's mayors and city managers watched as the first U.S. sale of a municipal wastewater treatment plant unfolded in the small town of Franklin, Ohio. Following the financial and operational success of this creative infrastructure deal, cash-strapped cities from around the country took notice and considered similar procurement strategies. The city of Wilmington, Delaware lead the way with its goal to be the first "large Franklin", initiating an asset sale valued at $50 million, almost ten times the Franklin deal.

Despite initial optimism, however, the Franklin model was not so easy to replicate, for a variety of reasons explored in this case. In addition to differences in scale, the Wilmington procurement involved political disagreement, lawsuits, revised federal regulations, and changing municipal finances, not encountered by the Franklin arrangement. The intended efficient transaction in Wilmington dragged on for nearly three years and emerged as a radically different agreement in the end. Nevertheless, Wilmington played an important role in developing options for other cities to follow. Its lessons heavily influenced the decisions of subsequent infrastructure procurements including water and wastewater plant negotiations in Cranston, RI, Hoboken, NJ and Taunton, MA.

Ellen Purlman's July, 1996 article in *Governing* magazine describes the motivation behind municipal interest in asset sales and long term operations contracts for waste water treatment plants:

> For municipalities stuck with the unglamorous and increasingly expensive business of owning and maintaining sewer systems, the Franklin deal looked like a problem-solver. It held out the hope that cities stuck with an aging plant that needed to be refreshed with new technology could get the facility brought up to date without bankrupting the municipality.

> What makes the Franklin option so exceptionally attractive is that the three cities and two counties that are served by the plant were able to pocket a profit of $1.8 million. Moreover, the deal leaves it to the private sector to invest new capital, modernize the facility and still keep a lid on user fees.

> Not surprisingly, other municipalities ... have been trying to follow in those footsteps. But for all the hoopla and excitement over the Franklin sale, what the Cranstons and Wilmingtons of this world are finding is that selling a wastewater treatment plant is no sure thing. Not that there aren't willing buyers. There are. But the path to a sewer sale is strewn with both the expected

fiscal and regulatory hurdles as well as surprisingly intense political problems.[33]

The Wilmington case raises fundamental questions about the strategic goals of any project delivery and finance strategy. Bluntly put, where in the hierarchy of public procurement planning should we rank quality of service, cost (to bidders, to users, to taxpayers), and power - i.e. to control the details of wastewater treatment? In an environment of uncharted public policy, what risks should a municipality take? The Wilmington case raises these questions, and more.

Wilmington, Delaware and surrounding New Castle County

Located approximately twenty-five miles south of Philadelphia, Pennsylvania, in northern New Castle County, the city of Wilmington, Delaware encompasses approximately sixteen square miles and a population of 71,500. Wilmington is the county seat of New Castle County and supports a wide range of industries, including many corporate headquarters for international companies. Favorable business regulations, including the absence of sales and inventory tax, have helped bring in a host of companies, though many maintain only token operations in Wilmington.

Like many U.S. metropolitan areas, the Wilmington region has experienced faster growth in the suburbs than in the central city limits. From 1960 to 1990, the affluent, suburban New Castle County grew from a population of 307,446 to 441946, a 44 percent change, while the central city shrank by 25 percent to 71,529 in the same period. The economics of the region have changed similarly. Tax revenues from both business and personal income grew steadily in the County. Meanwhile, revenues dwindled in the city though operating costs remained steady. These changes resulted in challenging economic circumstances for Wilmington in the late 1980's and early 1990's. Bond ratings deteriorated; payments on short term debt overshadowed long term planning; budget deficits grew to $10 million on a $120 million budget; and though some revenue funds were solvent, the city was in the red overall. (See Tables 13-1 and 13-2 for city and county budgets)

Exhibit 13-1   Water & Sewer Fund 1993, 1994

General Obligations Bonded Debt, June 30 ($000)	1994	1993
Various	54,856	54,883
Water & Sewer	55,049	54,957
Commerce	56,216	54,435
Home Ownership	5,000	5,000
Revenue Bonds	1,205	1,555
TOTAL	172,326	170,830
General Fund		
Taxes	48,350	45,108
Licenses	2,418	2,072
Fines, etc.	2,078	2,241
Interest	552	1,343
Other revenue	4,993	4,888
TOTAL REVENUE	**58,391**	**55,654**
General Government expenses	7,855	7,961
License and Inspection	1,310	1,337
Public Safety	28,618	27,692
Public Works	8,853	8,552
Municipal Court	1,180	1,251
Pensions	3,785	3,577
Recreation and Culture	4,020	3,923
Debt Service		
Principal	2,903	3,115
Interest	3,007	3,833
Other operating expenses	2,453	2,399
TOTAL EXPENDITURES	**63,984**	**63,640**
Excess expenditures	(5,592)	(7,986)
Rfdg. Proceeds	19,939	-
Escrow Payment	(19,937)	-
Transfers, net	cr: 2,630	cr 3,291
Begin Fund Balance	3,712	8,407
Ending Fund Balance	751	3,712

Table 13-1 City Budget
Source: Moody's Municipal & Government Manual, 1996, p.1336-1337, Moody's Investors Service, New York.

General Fund – Years ended June 30 ($000)	1995	1994
Property Taxes	66,060	62,127
+ Service Charges	11,040	11,706
+ Licenses and Permits	3,431	3,222
+ General & Administrative	3,327	3,261
+ Intergovernmental Revenues	3,909	3,258
+ Interest	3,448	2,069
+ Other revenues	2,002	1,241
TOTAL REVENUES	93,217	86,884
General expenses	16,018	15,841
- Public Safety	35,106	32,072
- Public Works	7,381	7,383
- Recreational & Cultural	9,305	8,791
- Judiciary Offices	4,702	4,565
- Libraries	4,722	4,264
- Debt Service	10,416	10,453
- Other Operating Expenses	145	149
TOTAL Expenditures	87,795	83,518
Excess Revenues	5,422	3,366
- Transfers, net	435	510
Beginning Fund Balance	23,023	20,167
Restatement of Beginning	3,717	
Ending Fund Balance	24,293	23,023

Table 13-2   New Castle County Budget
Source: Moody's Municipal & Government Manual, 1996, p. 1334, Moody's Investors Service, New York.

As the area around Wilmington developed, the city slowly lost its dominant position in controlling growth. Infrastructure - and associated costs, revenues, and decision-making authority - became an important issue for the city. Though the city owned and operated much of the regional infrastructure - including a wastewater treatment plant, water works, and port facilities - the majority of users and ratepayers were often county residents.

In this environment, interest in alternatives to owning and operating the wastewater treatment plant emerged.

Plant History

In 1995, the Wilmington plant was the sole wastewater treatment facility (WWTF) for nearly 50% of the residential population of Delaware, but it was owned and operated exclusively by the city. Operations entailed processing the wastewater from over 400,000 commercial and residential customers, including all 71,500 of users inside the city's limits. Usage was expected to grow with the population boom in the surrounding County.

The wastewater treatment facility was located within city limits along the Delaware River and included wastewater treatment operations (including primary clarifiers, polishing ponds, and chlorination tanks) and wastewater solids handling (including sludge holding tanks and solids dewatering processes). The process involved preliminary, primary, secondary, and tertiary treatment prior to release of the final effluent to the river. In addition, three sewage-pumping stations located throughout the city help feed influent to the plant.

Despite the economic challenges faced by the city, the water and wastewater treatment facilities had been operated successfully as self-sustaining enterprise funds for several years. The facility operated at a processing rate of over 90 million gallons per day (mgd) with an ultimate nominal hydraulic capacity of 350 mgd. The plant was slated for expansion to 105-mgd in the near future and the city had identified a number of projects that would enable this growth.

The wastewater treatment facility was constructed in stages from 1953 to 1992 at a cost of over $93 million dollars. Table 13-3 shows the source of these funds.

Source	Amount
City Funding	$37,680,237
State Grants	11,200,000
Federal Grants	44,900,000
TOTAL	93,780,237

Table 13-3   *Source of Funds for Construction of the Wilmington WWTF*

Like many cities, Wilmington built most of its wastewater treatment facility in the 1970s and 1980s, when federal funding pursuant to the Clean Water Act of 1972 (CWA) was plentiful. Under the CWA, which funded the EPA Construction Grants Program, the federal government paid seventy-five percent of the cost of design and construction. The remaining fifteen percent and ten percent were paid by state and local governments, respectively. Congress steadily reduced CWA funding throughout the 1980's and eventually cut off the previously guaranteed cash flow from these sources.

As the plant reached its design useful life, the aging infrastructure was in need of repair and renovation. Yet the city could no longer comfortably afford the initial capital expense of such projects. In 1988, a facilities plan was developed, describing necessary improvements to accommodate foreseeable growth in the plant through the year 2010. The initial phase included $30 million in improvements to primary treatment and solids handling.

Further improvements were in process, when the city decided to sell the facility. At that time, the main plant assets held a current book value (as of 12/31/94) of approximately $52.6 million and had been acquired at a cost of $88.2 million. Major components of the facility are described below in Table 13-4. Estimated replacement needs of the plant may be found in Table 13-5.

Asset	Acquisition Date	Acquisition Cost	Useful Life (years)	DEC 94 Accumulated Dep*	DEC 94 Book Value**
Sewage Treatment Plant	Jan 1, 1961	$3,643,753	32	$3,643,753	$0
Secondary Treatment Plant	Jan 1, 1966	$3,400,000	32	$3,145,456	254,544
Secondary Basins	Jan 1, 1976	$9,500,000	25	$7,101,048	$2,398,952
Screen / Chlorine Facility	Jan 1, 1977	$4,700,000	25	$3,656,254	$1,043,746
Sewage Treatment Plant	Jan 1, 1982	$28,794,331	32	$11,029,996	$17,764,335

* Depreciation is calculated using the straight-line method
** Book Value = Acquisition Cost - Accumulated Depreciation

Table 13-4	*Primary Plant Assets*
Source:	*Wilmington WWTP RFP Annex E. Page 1.*

Involvement of the Private Sector

Use of Private Contractors for Wastewater Treatment

Although Wilmington had a long history of private sector contracts for facility design and construction, the city first began examining the broader scope of operations and maintenance (O&M) contracts in the 1980's. At the time, contracts with the private sector were limited by Internal Revenue Service (IRS) regulations to at most five years terms. Beyond this period, tax-exempt bonds would be required to be defeased - a costly and time-consuming process. For this reason, O&M contracts typically revolved around short-term cost savings through work force reductions and more efficient management.   In many cities, private sector competition was sought for such routine services as trash collection, meter reading, and janitorial work.   Wilmington developed similar contracts at its wastewater facility for nearly a decade before discussing the sale of the plant.

ASSET (1)	Base 1995	FORECAST PERIOD (1)							TOTALS
		1996	1997	1998	1999	2000	2001-2005	2006-2015	
Pumping Stations:									
Replace screens								961	961
Replace main pumps				277	287			464	1,028
Miscellaneous			54				68	757	879
Prelim. Treatment:									
Replace screens			643				763	1,983	3,390
Replace grit classif.				499				1,633	2,131
Miscellaneous	150			166			212	587	965
Primary Treatment:									
Miscellaneous	150						212	568	779
Secondary Treatment:									
Fine bubble retrofit	600	621	643				2,371	4,211	7,846
Rehab sludge collector		621	643	665			2,454	3,122	7,506
Replace blowers			536				353	961	1,850
Polishing Ponds:									
Replace aerators		155							
Replace main pumps					230				
Miscellaneous					86				
Solids Handling:									
Digester covers	650	673	696						
Sludge heaters									
Replace belt presses					1,188				
Miscellaneous			107						
Other:									
Vehicles	70	16	75	17	23	18			
Odor/VOC controls	100	311		554		594			
Instrumentation/ SCADA		828							
MISCELLANEOUS	500				1,033				
TOTALS (SS)	2,220	3,224	3,396	2,179	1,658	1,799	11,460	31,968	55,684

Table 13-5   *Estimated Fixed Asset replacement Detail ($1,000's)*
            Source: Wilmington WWTP RFP, Appendix F

One of the larger wastewater contractors was Wheelabrator EOS (later acquired by U.S. Filter) that had first been selected as a private contractor-operator of the sludge dewatering facility in 1985. Good operating track records lead to its selection in 1989 for plant engineering and maintenance supervision services for the entire wastewater treatment facility.

Choice of an Asset Sale

The issuance of two Presidential Executive Orders in 1992 and 1994 for the first time allowed municipalities to sell portions of federally funded infrastructure to private industry as long as the main purpose of the facility or plant remained. This regulatory change motivated the precedent-setting Franklin decision and provided new alternatives to cities in managing infrastructure. This option also provided incentives for an infusion of private sector capital that did not exist with the previous five year O&M contracts. The change was particularly important for large plants such as wastewater treatment facilities, where high utility and plant maintenance costs could only be reduced through committed investment in automated controls and energy efficient equipment.

With this new option before it, the city of Wilmington began considering the sale of its wastewater treatment plant as one means of controlling costs while providing high quality wastewater treatment services to its ratepayers. Almost simultaneously, Wheelabrator's regional managers raised the subject with the city's water utility department manager. The idea interested the city and Wheelabrator made a series of presentations in March and August of 1994.

The city hired an outside consultant to investigate two different options for "privatizing" the plant. These options were described in the Request for Proposals (RFP) as follows:

> In October 1994, the City hired Raftelis Engineering Consulting Group (RECG) to assist in evaluating two privatization options: (1) leasing the WWTF to a private operator and (2) selling the plant to a private owner/operator. As compared to continued City

ownership and operation, the two options were evaluated to determine which provided the best opportunity for controlling WWTF operations costs; ensuring long-term cost stability; and generating a cash infusion for the city. Both options involved contract operations of the wastewater treatment facilities. The economic analysis included an evaluation of the short- and long-term impacts on rates, rate stability, and the distribution of economic risks. In addition, RECG analyzed how continued City ownership and operation, as well as the two privatization options, would affect non-economic criteria, including impacts on labor; operational risks; any loss of control by the city; quality of service; regulatory compliance; responsiveness to capital needs for expansion, rehabilitation, and facility upgrades; public acceptance; and ease of acceptance. The analysis determined that the sale of the WWTF assets as part of a professional services contract for management, operation, and maintenance provided the best opportunity for meeting the City's objectives and financial needs.

In January 1995, after reviewing the options, city officials decided to proceed with the procurement in the form of an asset sale.

Agreement with New Castle County

Another important aspect of the analysis was an evaluation of the impact of each alternative on the city's existing wastewater treatment agreement with New Castle County. Up to seventy percent of wastewater flows are generated by customers located in New Castle County. While the city provided treatment services to these County customers, the County maintained its own sewers and provided billing, collection, and customer service. The current City / County agreement obligated the city to provide wastewater treatment services for the County and included provisions for allocating wastewater treatment costs between the two bodies.

Shortly after, the city concluded to sell the plant, the decision was presented to county officials. The presentation to the county included possible outcomes and their effects on the service agreement. As Tom Noyes described in a 1998 interview: " The County didn't like it, but they did not say 'don't do it.'"

Issuance of the RFP

An RFP was issued on May 5, 1995, with the stated objective of forming "a public-private relationship with a private company for the ownership, management, operation, and maintenance of the Wilmington wastewater treatment plant and facilities." The successful proposer would be responsible for wastewater treatment operations, while the city would maintain the customer relationships. The RFP included the graphic shown in Exhibit 13-2 summarizing these responsibilities:

Exhibit 13-2   RFP Designated Responsibility Associations

The city's expectations for the asset sale and private operation included the following key points:

- Paying off relevant state and federal grants for the plant
- Defeasing outstanding debt
- Providing a cash infusion to the city to meet pressing financial needs
- Rate stability for its customers
- Continued high-quality delivery of treatment services
- Preservation of capital investment
- Access to external sources of capital through the private firm for expansion, upgrades, and replacement

- Employment opportunities to all present city employees with compensation, career advancement, and development opportunities comparable to or better than currently available

Financing of the Asset Sale

As described in an interview by Kash Srinivasan, the city's utility director, the swap of 20 years of operating revenues for upfront cash was like a home equity loan.[34] The asset sale consisted of the following main transactions:

The private party would

- make an up-front payment for the full value of the Purchased Assets (at Net Book Value), as determined at the time of the sale.

- produce proof of access to funding for the payment as well as future capital improvements.

The city would

- pay an annual service fee consisting of four parts: (a) a base component to cover operating costs plus a return; (b) a variable component; (c) a capital replacement and repair component; and (d) an Extraordinary Items Component.

- repurchase the facility for a pre-determined fixed price of $1 million 20 years later (This would be adjusted for the depreciated cost of interim additions).

The draft agreement consisted of three parts: a Professional Service Agreement, a Definitive Agreement for the Purchase and Sale of Assets, and a Ground Lease. Proposers were required to (1) satisfy the requirements of the agreement (2) define the costs to the City and (3) include additional insights to plant, process, and procurement options.

Competition

The Request for Proposals (RFP) attracted four competitors: Wheelabrator Water Services, American Anglian, U.S. Water, and Northern Delaware Clean Water Corp. (composed of Professional Services Group and Interwest Corporation, a non-profit taxpayer organized group). The competition was structured as a professional services contract, where procurement rules allowed the city to choose the winner on technical merit and then negotiate price with the most qualified bidder. Cost proposals would be kept secret, until after the price negotiations between the city and winning bidder were completed. In addition, each bidder was encouraged to provide a range of possible contract agreements, to include an asset sale as well as short-term and long-term operating leases. Although long-term leases were discouraged by IRS regulations at that time, the city wanted the additional proposal information to better understand the motivations and internal numbers of the private competitors.

This arrangement of open bidding was intended to lead to the best quality operations, while still allowing the city an advantage in negotiations. The bid sheet is shown in Table 13-6 below. In addition, the criteria for evaluating the technical qualifications of proposals are set forth in Table 13-7, also shown below.

In August 1995, Wheelabrator was chosen as the most technically qualified of the four competitors. Price negotiations started, but were soon interrupted by complications from the city.

Base Component Operating Costs	Contract Operations*	Sludge Dewatering	Biosolid Removal	Total
1. Personnel & Benefits				
2. Chemicals				
3. Annual M&R Budget				
4. Other Operating Costs				
Overhead/Fee				
Variable Component (Electricity Costs)				
Total Service Fee				

*Excluding Sludge Dewatering and Biosolids Removal

Table 13-6   *Service Fee Proposal Sheet*

Criteria	Points
Corporate Profile	8
Corporate Experience & Expertise	15
Regulatory Experience	15
Key Management and Operational Personnel	8
Financial Strength	15
Utilization of DBEs and EEO Compliance	10
Employee Considerations	12
References & Reputation	12
Completeness & Responsiveness of Proposal	5
TOTAL	100

Table 13-7   *Stated Evaluation Criteria*

In response, the four bidders submitted the following proposed transactions:

Bidder	Proposed Procurement Options
Wheelabrator	1. Base rate sale 2. Plant sale at an alternate valuation 3. Plant sale at an alternate valuation 4. 5-year operation service fee
American Anglian	1. Base rate sale 2. Sale at higher price 3. 20-year operating lease 4. Sale of a portion of plant
U.S. Water	1. Contract operations and management 2. Contract operations, including concession fee 3. Base rate sale 4. 15-year management contract
Northern Delaware Clean Water Corp.	1. Base rate sale by a non-profit and contract operations by a private company

Table 13-8   Responses to Wilmington RFP

Opposition to the Asset Sale

Suit by Northern Delaware Clean Water Corp.

In November 1995, one of the competitors, Northern Delaware Clean Water Corporation (NDCW), sued the city, claiming that the selection criteria incorrectly used professional services criteria for the procurement. The company claimed that the operation of a wastewater treatment plant is in fact a commodity service and therefore should be chosen by low bid only. NDCW's suit, if successful, could have forced the City to cancel the procurement and re-compete under different rules.

A Delaware court upheld the city's view that, as structured, the procurement sought a professional service. During the litigation, the base bids for all competitors were made public, as follows: American Anglian:

$13 million, Wheelabrator: $13.3 million, NDCW: $15 million, and U.S. Water $23 million. This compared to a city benchmark of $20 million.

Political Turmoil

The next obstacle arose in early 1996, in New Castle County, where opposition to the asset sale was growing. Between the first presentation a year early and current negotiations, county officials began to feel that the city intended to commit the county to increased rates, while benefiting from an exclusive up-front payment.

Irritated by this possibility, the county first responded by offering to buy the plant from the city, instead of allowing Wilmington to complete the deal with Wheelabrator. The city refused this offer and the political turmoil continued along with delays in the procurement. Finally, in June 1996, with the beginning of the pre-election season, the state Assembly passed a resolution prohibiting contract signing until the new city and county representatives were installed in January 1997.

Final Resolution: A 20-year Operations and Maintenance Concession

During the delays, important changes in regulations and political leadership took place. For one, the EPA loosened its interpretation of the Presidential Executive Orders and allowed more private capital investment in a plant without changing ownership from the city to the private contractor. Around the same time, treasury regulations were revisited concerning qualifications for tax exempt bonds in the case of long-term operation and investment by a private operator. The U.S. Conference of Mayors, along with specific mayors from major cities, lobbied for fewer restrictions, arguing that long term contracts - of 20 years or more - were essential in encouraging intelligent investment; private operators needed the time to write down costly initial investments in efficiency. The treasury department agreed and in January 1997 promulgated their decision to allow operating contracts for periods up to 20 years. The decision removed the complexities of having to sell the plant outright or defease municipal bonds in order to encourage long-term private capital investment.

Politically, tensions were easing between the city and county. Wilmington's mayor Sills was reelected with a high approval rating. Meanwhile a new leader who was more amenable to the city's proposal replaced the former County Executive. As long as the low annual rates could be guaranteed, Tom Gordon, the new County Executive was willing to allow the city to retain the plant and initiate a long-term contract.

In the same period as these important changes, the city found time to further analyze its options, learn more about the process, and re-assess its changing financial needs. Between 1994 and 1997, the city's general financial situation improved and the previous emphasis on an up-front cash flow from the sale of the plant became less critical. Instead, the city was able to focus on the potential annual cost savings. City management determined that a 20-year concession would provide greater long term economic benefit to the community than an outright sale - **with a present value difference of approximately $30-50 million, depending on how the analysis was done.**

The concession allowed the city to reap the operational cost savings of an asset sale, yet make the financial arrangements palatable to the county though lower service fees. With the twenty-year concession, the city retained ownership of the facility but allowed a private company to operate the plant at a profit as well as for the benefit of the public. The resulting comprehensive lease required the private operator to be responsible for utility, repair and tax expenses as well as normal operating costs. The city, however, retained ownership of the plant. This ownership, in the form of an enterprise fund, allows the city to include indirect costs and depreciation on the income statement and balance sheet. In addition, as described by Tom Noyes, executive assistant to the mayor, "ownership is control and cities these days don't control much of the infrastructure that is so important to them."

The final agreement was approved by the Wilmington city council in December 1997. Although it does not include an up front concession fee, it does lower operating costs, provide a $38-million capital component to be financed by U.S. Filter EOS, and set aside $1 million in transaction fees also to be paid by U.S. The service fee is expected to provide annual cost savings to the city and regional users of approximately three million dollars - or $60 million over the contract term.

Lessons Learned

The Wilmington competition was costly to both the public and private sides of the negotiations. Over the course of the three-year procurement, the city of Wilmington spent approximately $1million and U.S. Filter EOS spent at least $500,000. In 1997, Robert Joseph, EOS's project manager for the deal, maintained that despite the costs and delays, the contract still offered good returns for his company. Ownership of the asset was not the most important issue. Rather, a 20-year operations period and the ability to invest private capital early for long-term efficiencies were. He did, however, summarize his frustrations with the complicated procurement process:

> I feel like I've negotiated this contract not once, but four or five times. With this contract, the city could save about $10,000 per day, compared to existing operations of its aged facility. These savings are lost however because the contract signing is postponed each time by new issues. As time goes by, more unforeseen events are introduced. There really is a shelf life for this type of procurement.

Mike Gagliardo, the director of the Urban Water Council for the U.S. Conference of Mayors, noted that in hindsight Wilmington could have spent more time up-front working with the politicians, unions, and rate payers before going to the vendor community with an RFP. Nevertheless, he felt that the procurement strategy succeeded both in improving the city's infrastructure as well as providing lessons for other communities.

Looking back at the process in the spring of 1998, Tom Noyes was similarly happy with the outcome: "While I am sometimes frustrated that some cities started this process after us and completed it before, I do remind myself that we helped pave the way." Summing up his recommendations to fellow city officials considering public/private contracting options, he noted the following points:

1. Build flexibility into your procurement process from the start. Don't lock into only one model. The market provides options and the ability to switch to the best solution when more information or knowledge is available is valuable.

2. Look at other instances where cities tried similar procurement methods. Wilmington had few other role models.

3. Understand the politics behind the procurement.

4. Maintain a view of the big picture. Don't get bogged down only in the financials on the income statement and balance sheet. Also value the intangible benefits, such as the flexibility and control of maintaining ownership.

5. Check all assumptions made in preparing an economic model. Understand the effect of accounting changes.

Questions

1. Identify the "Client" that defined the project's scope – the City, the County, or both? What was the strategic goal of the procurement? Whose goal was it?

2. The City focused heavily upon a potential "sale" of the WWTF. Is the decision to sell or not to sell the central strategic issue? What should it have been?

3. Different options were included in the RFP for delivering these services. In which of the four quadrants does each option lie?

4. Mr. Noyes asserts that an RFP should include multiple delivery methods, i.e. scope need not be defined by the client before the

competition. Rob Joseph's perspective is almost exactly the opposite. Does the City's approach give the owner an unprincipled opportunity to select "who" it wants by selecting scope of work after proposals are submitted? Which, if any, of Mr. Noyes' recommendations preserve the City's long-term interests in focusing competition, attaining quality firms, and improving quality and cost performance?

5. Use the historical data in Tables 13-1 and 13-2 to project future sources and uses of funds for the General Fund and the Water & Sewer Fund (an enterprise account). Can you reliably predict future sources and uses in the General and Water and Sewer funds from this history? What information would improve the reliability of your prediction?

6. Evaluate the long-term contracting strategy as a means to convert unknown future City obligations into a stable, predictable, binding commitment by Wheelabrator? What are the tradeoffs?

7. Using the data in Tables 13-1 and 13-2 and historical cash flow analysis from question 5, construct a cash flow model for the final agreement with Wheelabrator, from the City's perspective.

References

Purlman, Ellen. "Selling off the Sewer." Governing Magazine July 1996, p 57.

"Private Plan in Wilmington." Engineering News Record March 20, 1995: 16.

Notes

---

[33] Ellen Purlman, "Selling off the Sewer," Governing Magazine, July 1996, p 57.

[34] "Private Plan in Wilmington," *Engineering News Record*, March 20, 1995, p. 16.

CHAPTER 14   THE STRATEGIC GOALS OF PUBLIC INFRASTRUCTURE PROCUREMENT: MASSACHUSETTS WATER RESOURCES AUTHORITY

Infrastructure Development Systems IDS-97-W-105

This case was prepared by Professor John B. Miller with the help of Research Assistant Maia A. Hansen as the basis for class discussion, and not to illustrate either effective or ineffective handling of infrastructure development related issues. Data presented in the case has been intentionally altered to simplify, focus, and to preserve individual confidentiality.

The Challenge of Infrastructure Investment in Wastewater Treatment

Clean water may seem like an obvious priority. The appearance of our cities, the health of our children and the condition of natural resources depend on clean waterways. Similarly, fishing, tourism, agricultural and real estate industries are all tied to the quality of water. Nevertheless, effective protection of water quality has been a daunting challenge.

Decades of regulation and federal funding have improved areas that were destroyed by unrestricted pollution, however much work remains. According to the U.S. Environmental Protection Agency's 1996 Clean Water Needs survey, $139.5 billion investment in publicly owned wastewater treatment facilities is still necessary to comply with Clean Water Act requirements through 2012. This includes $44 billion for wastewater treatment; $10.3 billion for upgrading existing wastewater collection systems; $21.6 for new sewer construction; and $44.7 billion for controlling combined sewer and storm water overflows.

Federal, state, and local regulations and funding were intended to salvage the nation's deteriorating waterways. The goal of the Clean Water

Act (CWA), the visionary program of 1972, was to eliminate the discharge of pollutants into navigable waters by 1985; to achieve a level of water quality by July 1, 1983, that protects fish, shellfish, and wildlife and lets them reproduce, and to stop the discharge of pollutants in toxic amounts. While progress has been made, these goals are yet to be met.

In this context, it is important to look back at the policies and specific cases that have succeeded and those that have failed to discover potential changes for future projects. In reviewing the impact of the CWA and others before it, it is instructive to look at how the resources were allocated through the procurement process.   While underestimated in the spirit of environmental legislation, procurement strategies have had an important impact on the overall success or failure of the program.

If our best lessons come from our mistakes, then Boston and the predecessors of the Massachusetts Water Resources Authority are a good place to start exploring the dynamics between procurement and the success of environmental policy. A notorious example of long-term environmental irresponsibility that influenced the 1988 Presidential Election, Boston is not alone in its challenges with water quality. The history of Boston's troubles with wastewater treatment - political, financial, and technical - mirror the challenges in managing large environmentally oriented infrastructure in the United States today.   Where procurement strategies revolve around the allocation of political control among numerous governments and agencies, the political maneuvering that follows can take priority over the real needs in local water quality.

Boston Harbor

Boston Harbor's history as the vital center of commercial and recreational activity for much of New England continues today in many forms. The Harbor encompasses 47 square miles and serves as a home for a large shellfish industry as well as the largest seaport in New England. It also supports numerous recreational activities such as sailing, swimming, and fishing.   Along its shores, the Harbor has attracted numerous commercial and residential developers, representing enormous economic investment in the area.

Despite its economic and recreational importance, the environmental health of the harbor has often been ignored. After three hundred years of being used as the city's water source and waste disposal site, Boston Harbor was seriously polluted from discharges of municipal and industrial wastes. In addition to being aesthetically unpleasant, these discharges lead to serious pollution problems such as eutrophication, concentrations of disease producing bacteria, and the accumulation of toxic substances in the water and shellfish. Harvesting of shellfish was prohibited in half the waters and more than 80 percent of the harbor finfish were diseased. (Table14-1 summarizes the history of the harbor) Despite federal attention, decades of funding, and public interest, the implementation of wastewater treatment has been met with varying levels of success.

In counterpoint to Boston, the city of Chicago provides a worthwhile comparison in procurement strategies. Both are older cities with similarly aging infrastructures. Both also treat large quantities of urban wastewater and discharge their treated effluent into bodies of water, which have received much attention for pollution in the past. Finally, both have spent billions of dollars to deal with increasing capacity and combined sewer / storm water outflows. Nevertheless, the two cities have implemented their solutions in completely different ways.

This case considers the policies that led to the recently completed Boston Harbor Project, one of the largest infrastructure projects in the United States. The case examines the project's history in relation to public policy at the time, and contrasts this with more modern approaches to public/private partnerships. The case raises broad issues about what local governments are best qualified to do, and how federal, state, and private sector organizations can complement these skills to create better solutions for the community and the environment.

Procurement and Regulatory Approaches to Wastewater Treatment - pre-1970

Like most coastal cities in the early years of this country, Boston's form of wastewater treatment entailed dumping raw sewage directly into the ocean or its tributaries. It was not until the late 1800s that the state of

Massachusetts first set up a regional sewer commission, which took the first step of screening large debris from the wastewater.

The early part of the 20th century was marked by sporadic, locally funded investment in wastewater treatment infrastructure. Three treatment plants were constructed in the Boston area and then consolidated under the newly formed Metropolitan District Commission (MDC). While a start, these measures were insufficient. By the 1920's, less than 30% of urban wastewater in the U.S. was treated, and little of that was highly effective. Increasing discharges of municipal and industrial wastes left poor water quality in such dense urban areas as New York and Boston.

Congress reacted by passing the Water Pollution Control Act of 1948, which for the first time, authorized federal funds for state and local water quality programs. This act included a grant program to help local governments design and plan wastewater treatment plants, but Congress never appropriated any money for construction.

Construction was not addressed federally until Congress passed the Federal Water Pollution Control Act (FWPCA) of 1956 that created the construction grants program, the precursor to the Clean Water Act. Through this program, local governments could receive 30% of the estimated amount of plant construction costs, or up to $250,000.

Further federal oversight followed with regulations in 1965 requiring states to develop water quality standards for interstate waters. This act also created the Federal Water Pollution Control Administration to establish broad guidelines, approve state standards, and assume responsibility for administering federal assistance.

During this period, Boston built the primary treatment plants at Nut Island (1952) and Deer Island (1967). Nevertheless, the limited primary treatment process suffered from serious flaws which led to the discharge of raw or partially treated sewage into Boston Harbor and its adjacent waters; 500 million gallons of partially treated effluent and 500,000 gallons of toxic sludge were discharged daily into the harbor in the outgoing tide. Old designs had intentionally combined sewer and storm drains into the same collection and treatment system. During rainstorms, the surging flow of

Water Distribution and Treatment

1652: "Water Works Company" incorporated, build first cistern and water conduit

1795: First network bringing water from beyond City limits (Jamaica Pond)

1848 - 1898: Network spreads to bring water from Lake Cochituate, Sudbury River

1895: Metropolitan Water District created (10 mile radius of State House), major expansion

1919 - 1970: continued expansion and increase in demand.

1980: Water conservation and demand management begun - concerns that demand will increase dramatically in future

Current Situation: New treatment requirements for the 1986 amendments Safe Drinking Water Act (cost of complying $560 million, asking for waiver)

Debate with Commonwealth, MWRA, and MDC Watershed Division over how much MDC should charge MWRA for water at source

The Sewer System

1885: Boston Main Drainage System originally placed in operation. Some original components still in use

1889: Metropolitan Sewerage District created - sewerage pumped to remote storage tanks and then released into ocean.

1952: First primary wastewater treatment plant constructed at Nut Island, in response to concerns about water pollution.

1967: Water Treatment Plant at Deer Island completed

1982: Litigation commenced by the city of Quincy against MDC to halt pollution of Quincy Bay and other areas of Boston Harbor.

1983: Litigation commenced by Conservation Law Foundation of New England.

1985: EPA names MWRA in violation of Clean Water Act

Current Situation: Authority serves 43 communities with 5400 miles of local sewers

System divided into two sections:

Northern system serves 1.26 million people and pumps wastewater to Deer Island (designed for average daily flow of 343 mgd and peak of 848 mgd)

Southern system serves 700,000 and pumps to Nut Island (designed for average daily flow of 112 mgd and peak of 290 mgd)

Sewers (built mostly prior to 1910) were designed to combine wastewater and storm water overflows - during rains this leads to discharge of raw water and significant source of pollution into Boston Harbor

Sewer pipes are old and leaking

Capital Improvements Planning

(for construction, operations, and maintenance of capital facilities):

Rolling three-year planning periods

Capital Improvements from 1985 to 1991 totaled $800 million.

1992-1994 program: $6.3 billion in planned projects ($4.7 for wastewater, $1.3 water)

*Deer Island Harbor Project: largest project in CIP, largest public works project ever undertaken in New England. Second largest sewage treatment plant in nation. Design and Construction of new primary and secondary wastewater treatment facilities with capacity up to 1.3 billion gallons per day from 43 communities.

Table 14-1   *Boston's History with Water Treatment*

water would exceed the treatment capacity, causing untreated water to pour into the harbor. These discharges made the Boston Harbor among the most polluted bodies of water in the nation.

The Clean Water Act

Like Boston Harbor, the quality of national waterways continued to decline, despite the regulatory action of the 1940s through 60s. By the 1970's, only a third of the nation's waters were safe for fishing and swimming, and sewage treatment plants served only 85 million people. Growing citizen concern called for more governmental involvement. Heated debates about the role of federal policy and spending in the health of local waters continued, until 1972.

On October 18, 1972, Congress overcame a presidential veto and passed the Federal Water Pollution Control Act Amendment, commonly known as the Clean Water Act (CWA) to "restore and maintain the chemical, physical, and biological integrity of the nation's waters." The law created the National Pollutant Discharge Elimination System (NPDES) permit program, which required states to establish total maximum daily loads for pollutants in their waters and best available technology standards for industry. It also required an upgrade of municipal wastewater facilities from primary to secondary treatment, created a national pretreatment program, and authorized $18 billion in grants for municipal wastewater treatment infrastructure over a 3-year period.

CWA was modified in 1977, 1981, and 1987. The 1977 amendment encouraged states to manage the NPDES and Construction Grants programs. The 1981 statute reduced federal funding for wastewater treatment infrastructure. In the four years after CWA was enacted, Congress appropriated a total of $18 billion for the program. As a result, the number of wastewater plants with secondary treatment grew by more than 58% from 1976 to 1980. Funding for the program began to decline in the late 1970s and averaged about $2 billion annually in the 1980s.

When the Reagan administration made it clear that federal grants were no longer a priority for wastewater treatment, Congress replaced the

Construction Grants Program with the State Revolving Fund (SRF) Program, which provides low-cost sources of financing to cities for environmental infrastructure needs. Under the program, the federal government provides capitalization grants that cover 83% of a state's request. States provide the difference and offer a range of low-interest loans to local governments. The fund is then replenished as loans are repaid. From FY 1989 to FY 1997, the SRF program provided states with nearly $13 billion.

Boston's Response to the Clean Water Act

Boston's reaction to the offer of federal funds was capricious and ultimately extremely costly. The wastewater system was managed by the Metropolitan District Commission (MDC), yet each of the surrounding 43 communities maintained an independent local sewer collection system that fed into 228 miles of MDC interceptor sewers. Many parties were involved, and although the need for improved treatment was easy to see, political entities could not agree on a solution.

Meanwhile, federal regulations pursuant to the Clean Water Act provided 75% of the funding for design and construction of wastewater treatment plants to municipalities on a first come first serve basis. This was matched by 15% from state funds, requiring only a minor investment on the part of the city. The result was an incentive for municipalities to over-build locally. This developed into a network of excessive capacity in areas that were proactive and under-capacity in areas which procrastinated.

Boston was a prime example of the latter. While the bickering continued between local Boston groups in the 1970's and 80's, the needs of the water system went unmet. Boston never applied for the construction grants, and instead relied on waivers to stay the investment. Boston, as well as many other cities, applied for special waivers but was denied. Between the applications for waivers and local disagreements about funding and placement of a new treatment site, federal funding had dried up.

Creation of the MWRA

In response to increasing pollution, shellfish contamination and closed beaches, the residents of the coastal town of Quincy filed suit against the MDC for its mismanagement of the wastewater treatment system. As a direct result, the MWRA was created by the Massachusetts legislature and in 1984 assumed possession and control of water and sewer systems from the floundering MDC.

In order to decouple political decisions from infrastructure investments, the MWRA was established as an independent authority supported by a base of wholesale user charges. The MWRA's powers were essential to get the harbor cleanup underway and to unfetter the program from regional squabbles in the state house. Based on years of experience, local politicians were found to defer to ratepayer price pressure, rather than raise rates and invest in needed treatment plants and distribution systems. Lacking federal funds for the project, this independent power was seen as the only sustainable method of forcing the constituents to fund the costly but mandatory harbor clean up.

In this light, the authority serves as the wholesale distributor of water and sewer services to the surrounding districts and municipalities. See Figure 14-1. Local governments, in turn, are the retailers and bill collectors for the 2.6 million people in the service region. Should local residents or municipalities fail to pay their bills, the authority is empowered to intercept (without suit) state aid funds earmarked for the localities. Unlike other state organizations, the MWRA is not subject to limits on charges to its users. The MWRA is required by the Act to set its rates at levels sufficient to pay its current expenses and debt service.

Even this arrangement was not enough. For years, in fact, the MWRA required oversight by a state or federal level judge. The courts set and enforced milestones that provided the structure and threat of fines to force compliance with the CWA. Boston would be forced to clean up the harbor to meet the requirements of the suit, but it would have to do the work at the expense of local ratepayers.

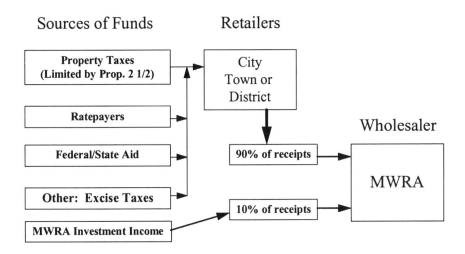

Figure 14-1   Channels of Payment

In 1997, the MWRA had an extensive and ambitious schedule of capital improvements, mostly subject to deadlines as set forth in orders of the United States District Court in the Clean Water Act Enforcement. The Authority was empowered, independent of regular legislative approval, to issue revenue bonds and notes, hire personnel, adopt budgets (after public hearing), expand service areas, acquire property, lease and sell property, and establish rates and charges.

Summarizing the problems that Boston faced, U.S. News and World Report noted in 1990:

EXPENSES	Sewer	Water	TOTAL
Operating & Maintenance Costs			
Direct Expenses	100,258,844	45,396,153	145,654,997
+			0
Other Costs:			0
Allocated Support Division Expenses	29,481,185	13,245,170	42,726,355
Operating Reserves	0	2,115,000	2,115,000
Other Allocated Indirect Expenses	10,111,952	968,303	11,080,255
+ (PLUS)			0
Capital Costs:			0
Debt Service (less State Debt Service Assistance)	148,923,610	14,310,966	163,234,576
Renewal & Replacement Reserve	0	0	0
Construction Cash Financing	916,458	0	916,458
- (LESS)			0
NON-RATE REVENUE	(30,050,000)	(6,717,000)	(36,767,000)
Investment Income	0	(73,448)	(73,448)
Fees	(2,670,725)	(3,701,382)	(6,372,107)
Miscellaneous Revenue	(600,000)	0	(600,000)
Rate Stabilization			0
Equals =			0
RATE REVENUE REQUIREMENT	256,371,324	65,543,762	321,915,086

Table 14-2   Calculation of Rate Requirements for FY 1997
Source: MWRA FY 1997 Current Expense Budget, Page 134

Allowing the municipal infrastructure to decay, as Bostonians have learned, is fiscal folly. If today's sewer modernization had been tackled in the mid-70's, it would have cost perhaps $900 million, most of it financed by federal grants. But mired in political bickering and public indifference, authorities from Governor Dukakis on down spent a decade ignoring Boston's leaky, decrepit sewer system. Twice they filed for waivers from the federal cleanup laws. Many other cities, meanwhile, were getting from 70 to 90 percent financing from the U.S. Environmental Protection Agency, which distributed almost $60 billion for sewer modernization over the past 15 years. Today, however, the flow of federal dollars has dried to a trickle, and next year's White House budget request, to be spread nationwide, is only $1.6 billion. This leaves Boston-area residents no choice but to foot almost the entire cleanup bill itself.[35]

Exhibit 14-1, located on the companion CD, provides a spreadsheet to examine the cost of Deer Island's construction, in total, as well as its direct cost to rate payers. Details of MWRA's financial condition are presented in Exhibit 14-2, also on the companion CD. Table 14-3 below shows details of the budget for the North and South Sewer System, which includes the Deer Island plant and accounts for over 40% of the sewer division budget.

Exhibit 14-1   *Framework for Analyzing Funding for Deer Island*
Exhibit 14-2   *Summary of MWRA Financial Condition*

Current Expense Budget, FY 1997
North and South System Wastewater
Treatment Processes ($000s)

	FY95	FY96	FY97
Line Item	*Actual*	*Actual (Unaudited)*	*Budget*
Wage & Salaries	12,366	15,744	16,826
Overtime	1,582	1,921	1,750
Fringe Benefits	43	63	35
Chemicals	6,304	6,071	7,114
Utilities	3,110	6,524	7,118
Maintenance	4,063	3,372	4,678
Training & Meetings	75	66	40
Professional Services	316	1,365	1,441
Other Materials	883	1,071	595
Other Services	974	409	540
TOTAL	29,716	36,604	40,136

Table 14-3 MWRA Operational Budget for the Sewer Division
Source: MWRA FY 1997 Current Expense Budget, page III-36

Description of the Harbor Clean-up

The $6.1 billion Harbor Clean-up project, the largest wastewater project in U.S. history and the second largest public works project ever undertaken in New England, is a result of Boston's procrastination. It will include primary and secondary treatment plants, egg-shaped sludge digestors, a sludge processing plant and two undersea tunnels. The centerpiece of the project is the construction of a new primary and secondary treatment plant on Deer Island. Once completed, the $3.25 billion plant will become the second-largest sewage treatment plant in the nation.

Construction of the new primary plant began in 1990 and the MWRA began operation of the first phase on schedule in 1995. Construction of the secondary plant began in late 1992. The new Deer Island plant provides state-of-the-art primary treatment of 1.9 billion gallons per day, which will remove 55 percent of solids and reduce biochemical oxygen demand (BOD) by 35 percent. Secondary treatment will further purify the wastewater, removing 85 percent of solids and 50 to 90 percent of toxins. BOD will be reduced finally to a total of 85 percent. The project is expected to be constructed under 80 contracts, with 15 phases.

Because of the high construction and debt service costs, the Authority's rates and charges, already among the highest in the nation, will increase significantly each year in response to continuing increases in the operating expenses of the systems. From 1985 to 1993, aggregate rates increased by more than 425%, to $243. By FY 1997, the rates were expected to reach $754. Exhibit 14-3 shows annual water and sewer charges from 1991-1996. The steep increases in rates, dictated in large part by the court-ordered portions of the Capital Program, will undoubtedly have substantial financial impacts on the local bodies, possibly generating public opposition to the programs.   To finance the construction, the authority would become the most prolific U.S. issuer of municipal revenue bonds through the 1990's.

Exhibit 14-3   Combined Annual Water & Sewer Charges in MWRA Communities

Table 14-4 shows expected MWRA wholesale charges to many of the MWRA's member communities.

MWRA Community	Total O&M Charge	Total Capital Charge	Total Charge
Arlington	2,418,086	2,687,489	5,105,575
Ashland	296,491	572,299	868,790
Bedford	1,210,311	825,972	2,036,283
Belmont	1,783,274	1,580,153	3,363,427
BWSC	40,542,097	35,208,136	75,750,233
Braintree	2,650,938	2,202,946	4,853,884
Brookline	4,606,901	3,684,555	8,291,456
Burlington	1,372,166	1,398,770	2,770,936
Cambridge	7,189,775	6,439,958	13,629,733
Canton	1,643,079	1,194,222	2,837,301
Chelsea	1,874,832	1,642,813	3,517,645
Dedham	2,521,974	1,686,798	4,208,772
Everett	2,457,492	2,186,814	4,644,306
Framingham	4,034,800	3,842,892	7,877,692
Hingham	566,589	408,552	975,141
Holbrook	325,994	505,889	831,883
Lexington	2,685,255	1,996,655	4,681,910
Malden	3,657,578	3,317,053	6,974,631
Medford	3,689,820	3,534,521	7,224,341
Melrose	1,762,516	1,766,653	3,529,169
Milton	1,855,657	1,627,869	3,483,526
Natick	1,868,452	1,779,351	3,647,803
Needham	2,307,322	1,842,126	4,149,448
Newton	8,848,402	6,097,316	14,945,718
Norwood	2,654,521	2,020,991	4,675,512
Quincy	6,259,961	5,424,257	11,684,218
Randolph	1,610,591	1,823,088	3,433,679
Reading	1,355,666	1,382,473	2,738,139
Revere	3,059,326	2,723,695	5,783,021
Somerville	4,012,231	4,245,047	8,257,278
Stoneham	1,931,902	1,485,036	3,416,938
Stoughton	1,421,340	1,376,362	2,797,702
Wakefield	2,055,451	1,691,050	3,746,501
Walpole	1,032,849	1,080,888	2,113,737

Table 14-4   1998 Wholesale Charges
Source: MWRA – Preliminary FY 1998

Exhibit 14-4, included on the companion CD, shows a graph of comparative annual rates in other U.S. cities.

Exhibit 14-4   Sewer and Combined Charges, 1996

Comparison to Chicago TARP Project

Boston is not alone in its poor history in addressing infrastructure needs and applying for available federal grants. Chicago, however, was one of the many who invested in wastewater fixes early on. As a result, it has been able to take advantage of federal funds available through the U.S. EPA's construction grants program.

While Boston's Deer Island is expected to be the second largest treatment plant in the U.S., Chicago's combined sewer/stormwater run-off project is the largest. Planning started for Chicago's Tunnel and Reservoir Project (TARP) in the 1950's and construction started in 1976. Work is not expected to reach completion until 2003 and the expected total cost is expected to reach nearly $4 billion. The project itself involves a giant 1 billion gallon per day pumphouse and 131 miles of tunnels, each up to 33 feet wide. The system includes three large reservoirs that collectively will hold more than 40 billion gallons of storm and sewage runoff. The goal of the system is to hold runoff during storms until it can be treated and released into Lake Michigan. At the project's half-way point in 1993, sewer overflows and basement flooding had been reduced by 75% and water quality at local beaches had improved enough so that they could remain open after heavy rains.

TARP's owner, the Metropolitan Sanitary District of Greater Chicago (MSD) secured 75% of the $1.2 billion construction cost of phase one from the Environmental Protection Agency's construction grants program. The remainder came from local taxes. In comparison, if Massachusetts had started its construction program in the early 1970's, 40% of the project could have been built with federal funds, according to an official with the Metropolitan Water Reclamation District of Greater Chicago.

Increased Private Sector Involvement from the 1980's

In the 1980's and 90's, cities around the nation began wrestling with the high operating and maintenance costs required for plants built soon after the CWA. The installed base of water and wastewater treatment plants is large and spread out. Most of these plants were built in the 1970's and 80's and are currently in need of over $100 billion worth of maintenance and upgrades. In addition, legislatures and utility managers have yet to find the most effective way to fund these expensive projects. Currently, funding is limited to much smaller State Revolving Loan Funds. The future is complicated, but a resolution is necessary. Procurement frameworks for these large projects will play crucial roles in completing the work.

	Installed Base	%Publicly Owned	Total Annual Revenues
Municipal Water Works	over 55,000	80%	$25.3 billion
Publicly Owned Treatment Works	27,000	95%	$27.3 billion

Table 14-5   Installed Base of Water and Wastewater Systems
Source: Compiled by case writer from EPA web page

Treatment Category	Investment Needs ($ billion)
Secondary Treatment	26.5
Advanced Treatment	17.5
Infiltration/Inflow Correction	3.3
Replacement / Rehabilitation	7.0
New Collector Sewers	10.8
New Interceptor Sewers	10.8
Combined Sewer Outflows	44.7
Stormwater	7.4
Non-point Source	9.4
Urban Runoff	1.1
Groundwater, Estuaries & Wetlands	1.1
Total	139.5

Table 14-6   Summary of Needed Investment
Source: 1996 EPA Needs Assessment for Publicly Owned Treatment Works

As the construction of the Deer Island Plant neared its end, the MWRA began to prepare for the challenges of efficiently managing the new system. The MWRA began studying the actions of other large cities and actively considering the latest trends in the wastewater industry. One of these options is increasing the use of a public-private structure for many types of work. In the water/wastewater business in general and at MWRA in particular, the private sector is usually engaged in the design and construction of utility assets. In addition, non-core maintenance activities have often been outsourced. MWRA used private contractors to design and construct the plants and currently outsources such activities as security services, laboratory testing, and janitorial services.

The MWRA briefly considered the option of private contract operation of the new Deer Island plants. The authority's views were expressed in a 1997 MWRA Briefing Paper on Private Sector Roles in Operations and Maintenance:

> No contract O&M proposal for a wastewater treatment plant nearly as large or complicated as Deer Island has been advanced anywhere in the country. Undertaking such a process would be a major commitment requiring efforts of staff as well as specialized consultants that would be far beyond the scope of effort that any other community or system has embarked upon. ... Millions of dollars have been invested in training of plant staff to perform and support the turnover / start-up exercise and hundreds of thousands of hours has built the skill and experience that today supports the ongoing start-up effort. To displace the current team operating and maintaining the plant mid-way through the start-up process would require a very elaborate transition exercise at the plant.

Additional analysis by the MWRA with respect to public/private contracting is included on the companion CD as Exhibit 14-5.

Exhibit 14-5   MWRA analysis of Contract O&M Issues

Based on MWRA's analysis of its plant specifics and the experiences in the industry, the management decided to continue examining outsourcing opportunities in smaller areas of their operations. They committed to develop a framework for "managed competition" where current employees can compete with the private sector for contracts. Further study of contract O&M for smaller plants was recommended, but the Deer Island plant, for the moment, would remain publicly run.

A number of cities - both large and small - are not as reluctant to encourage private sector involvement. Several have recently completed procurement strategies that actively involve the private sector in operating the majority of their wastewater treatment facility. A summary of the deals as of early 1998 is shown in Exhibit 14-6, on the companion CD.

Exhibit 14-6   Public/Private Partnerships

In 1996, the U.S. Conference of Mayors conducted a survey of 261 cities and found that 39% of cities have some form of private sector involvement in their wastewater services. Of those that did not currently have private sector involvement, 14% said they were considering forming a water partnership and 11% were considering forming a wastewater partnership.

The most significant incentive to implementing public/private partnerships was to reduce costs. Local issues - specifically labor barriers, intergovernmental relations and procurement restriction - were identified as the biggest impediments.

Questions

1. The political difficulties associated with the Wilmington project are trivial compared to those in the Boston case. Did the MDC's historical orientation as a regional water authority, with a widely dispersed collection of physical assets, fit well into the Clean Water Act's funding solution? What assumptions does the Clean Water Act make about who the Clients would be?

2. Thoedore Lowi argues, in "The End of Liberalism", that governments cannot plan under these circumstances. Can multiple governments effectively plan an integrated approach to water delivery and finance in circumstances like Boston?

3. Use the information in the case to test Boston's alternatives at each stage.

4. How should the Clean Water Act have been structured to avoid problems discussed in this case? Is a strategy that relies on multiple governments acting independently in Quadrant IV workable? Will it be workable in the future?

References

"The Boston Harbor Mess." U.S. News & World Report September 24, 1990: 58.

Notes

_____

[35] " The Boston Harbor Mess", U.S. News & World Report, September 24, 1990, p. 58.

CHAPTER 15   TOLT RIVER WATER TREATMENT PROJECT

Infrastructure Development Systems IDS-98-W-106

Research Assistants Bradley Moriarty and Katie Adams prepared this case under the supervision of Professor John B. Miller as the basis for class discussion, and not to illustrate either effective or ineffective handling of infrastructure development related issues. Data presented in the case has been altered to simplify, focus, and to preserve individual confidentiality. The assistance of Elizabeth S. Kelly from the City of Seattle, and Paul R. Brown of CDM Philip in the preparation of this case is gratefully acknowledged.

Introduction

During the 1996 gubernatorial race in the State of Washington, considerable focus was given to the issue of public infrastructure spending. The winner of the race, Gary Locke, pledged to allow cities to pursue alternative delivery strategies in order to reduce public spending without reducing the amount of completed infrastructure projects. His solution was to propose a bill to the legislature that legalized Design-Build (DB), Design-Build-Operate (DBO), and Build-Operate-Transfer (BOT) procurement strategies for public projects within the state.

Deborah Keller is an Administrative Assistant in the governor's office. Since her undergraduate work was in Civil Engineering, she was given the task of researching the practicality of a statewide bill and its impacts on communities in Washington.

Deborah chose the City of Seattle as the starting point of her investigation because of their recent DBO procurement of the Tolt Water Treatment Plant.

Project Background

The water supply for the City of Seattle and the surrounding counties, King and Snohomish, flows from three sources: the Cedar River in southeast King County, the Tolt River in northeast King County, and a well field in the Highline area to the south of central Seattle. Cedar River provides 66% of the water supply and the Tolt River provides about 28% of the water supply, with the Highline well field contributing the remaining 6%.

Exhibit 15-1   *Seattle Metro Map*
Exhibit 15-2   *Seattle PU Location Map*

The problem for the City of Seattle lay with the Tolt water supply. The River and its reservoir are subject to great swings in turbidity. The Tolt River receives increased runoff from the watershed during the rainy winter months. This runoff dramatically increases the turbidity of the source, which has caused it to be shut off as a raw water source nearly every other year for extended periods during the winter season. In the summer months, when water usage reached its peak, the Tolt reservoir was drawn down to help meet the need for raw water. As the reservoir was brought to lower levels, the wind and wave action increased turbidity and decreased the usable volume of the reservoir. Because of these turbidity fluctuations, the flexibility of the whole water supply system was limited to the capacity of the Cedar River source and the Highline well field.

Exhibit 15-3   *SPU Vicinity Map*

The Design Build Operate Decision

In response to increasing water needs in the metro Seattle area, the city identified the need to improve the predictability of the Tolt River water supply while complying with current and future health and safety standards and allowing for expansion to meet the future needs of the City and its environs. On December 8, 1993 Council Resolution #29251 was adopted allowing the water department to explore the Design-Build-Operate (DBO) option for the procurement of the Tolt Water Treatment Plant (see Exhibit 15-4 on the companion CD). Further resolutions and ordinances were adopted to allow the water department to enter into contracts with consultants and contractors. The most significant of these arrangements was the retention of the joint venture of RW Beck/Malcom Pirine to serve as the City's independent checking engineer throughout the procurement (see Exhibits 15-5 and 15-6 on the companion CD). These resolutions were necessary because the City of Seattle follows the federal guidelines for infrastructure procurement, which only allows the Design-Bid-Build (DBB) strategy to be implemented.

Exhibit 15-4	*Resolution and Ordinances*
Exhibit 15-5	*Benchmark Spreadsheet*
Exhibit 15-6	*Benchmark Drawings*

The City Council chose to investigate DBO because of the predicted increase in efficiency in the construction and operating processes (on the order of 10 to 15%), owing to collaboration among the designer, contractor, and operator. DBO also has the added attraction of further removing the city from the task of policing disputes between designer and contractor. This strategy would allow the city to fix costs for the project earlier in the design procurement process and also give the design/contractor the ability to "fast track" where critical components can be purchased prior to final design completion.[36]

Efficiencies would also be achieved on the operations side. A DBO joint venture would provide incentives to design and construct the facility for minimum maintenance costs and maximum efficiency. This maintenance company, assuming their expertise comes from operation of many such projects, would also be able to benefit from economies of scale.

The City also hoped that the international trend towards DBO and other long-term contractual agreements would provide incentive for companies to bid very competitively in an effort to be on the forefront of the market.

Objectives and Scope

In the Background Issue Paper as Prepared for City Council, November 1995, the objectives for the implementation approach were outlined as follows: (See Exhibit 15-7 on the companion CD for the entire paper).

Exhibit 15-7   Background Issue Paper as Prepared for City Council

"...the department expects, under any implementation approach selected, to maintain project ownership and responsibility for:

- Overall project management.
- Project financing.
- Specification of minimum conditions for environmental permitting and
  mitigation.
- Site access, consisting of the roads and bridges between the gate at Kelly Road and the Regulating Basin, and
- Ongoing public health protection and regulatory responsibilities.

Specific project objectives are added in several phases:

Implementation process objectives are to:
- Assure a fair, open market solicitation process.
- Protect the City's interests, and
- Allocate risk for project implementation to those parties (private and public) best suited to protect the public interest.

Design and construction objectives are to assure:
- Optimization of present and future water treatment processes.
- Efficient environmental permitting and mitigation.
- Aggressive scheduling.
- Low construction costs without overruns, and
- A high degree of design/build coordination at minimal risk to the City.

Plant operation & maintenance objectives are to assure:
- Reliable, efficient water treatment services to the public.
- Ongoing compliance with all applicable regulations.
- Effective response to both standard and unusual operating situations, and
- the lowest possible operational cost (to) rate payers, and
- prudent management and protection of public resources."

Benchmarking

Seattle Public Utilities (SPU) used special ordinance 118008 to engage a group of consulting engineers from R.W. Beck, Inc., Malcolm Pirnie, and Moore, Culp and Raftelis to investigate the costs and methods to design, construct, and operate the plant. The team served as an independent reviewing team to assist the City in the development of an RFP that clearly described the scope of services solicited by the City and the system by which proposals would be evaluated. The independent reviewing team accomplished the ends by preparing a benchmark design for the project that was included in the RFP. From this investigation, they created a benchmark design for the proposal. SPU negotiated a fixed price of $643,481.00 for these services.

The Benchmark was essentially the first step in the traditional Design-Bid-Build strategy. It included all of the appropriate elements for the project, as well as the designer's estimate of price for the design, construction, and 25-year operation of the plant. This Benchmark and its final price formed the mechanical standard and the price baseline for all the proposals.

The costs for construction, operations, and maintenance were broken into several capital and operating elements. The capital cost elements were site permitting, engineering design, construction management services, site work, facility construction, and financing costs, while the operating cost elements were labor, power, supplies, chemicals, maintenance, and equipment replacement. The Benchmark estimate included the estimated prices per unit for each of these elements. Once the entire project cost was estimated, the construction and operation and maintenance forecasts were converted to a Net Present Value (NPV) at the rate the City expected to be charged for the City's tax exempt bonds, 6.09%. By converting the financial model of the project from a long cash flow into an NPV (and requiring the proposers do the same for their proposals at the same discount rate) SPU sought to assure that all proposals were compared with the same method.

Because the cost of power, labor, chemicals and other goods and services will change over the lifetime of the project, SPU created indices for adjusting prices according to the market demand. In their final agreement with CDM Philip, these indices were called the "Construction Price Index," and the "Operations Price Index." The indices were both set at 3% for purposes of calculating the Benchmark.

In the Benchmark, Beck/Pirnie assumed that design and construction would take four years and would begin in January 1997. That would be followed by 25 years of operations before the facility was turned over to SPU. Yearly drawdown caps would finance construction as follows:

Year	Amount (1998 $$)
1997	$5,121,000
1998	$29,691,000
1999	$36,863,000
2000	$30,719,000
NPV	$101,000,000

Table 15-1   Yearly Drawdown Caps

The total cost of the Benchmark included the NPV of these drawdowns (the Fixed Construction Price) plus the NPV of the 25 years of Service Fees paid to operate the facility.

Construction NPV= $101,000,000
Operation   NPV= $ 55,900,000
Total   = $156,900,000

85% of Total   = $133,365,000

In the DBB strategy, the fixed construction price will be distributed according to the following schedule:

Year	Amount (%)
1997	0
1998	27
1999	40
2000	33

Table 15-2   Fixed Construction Price Distribution Schedule

The remainder of payments will be made in the form of a yearly Service Fee payable in monthly installments. The Service Fee includes a general fee for the operating company, a fee for chemicals, a "pass through" fee (composed of fixed costs such as Gas, Electric and Phone Service), an incentive/adjustment fee (to account for unforeseen occurrences), and funds to be paid for periodic repairs and replacements as per the negotiated schedule.

In effect, the Benchmark solidified the City's cash flow commitments to specific payments at specific times, a significant advantage to cities and towns with complex capital budgeting problems.

The Bid Requirements

SPU stipulated that to be responsive (i.e. considered), proposals could be no more than 85% of the Benchmark, $133,365,000. While this guaranteed that SPU would save money over the traditional DBB approach, this required that the City understand the proposals in far greater detail than would otherwise be required to determine the feasibility and comprehensiveness of the proposals. The Benchmark gave SPU a clear picture of the scope and costs involved in a traditional procurement system, and also provided the proposal teams with a catalog of SPU's expectations. While this allowed SPU to set the standard, it also opened up the field to novel solutions and designs, allowing any design that provided water at the required rate and clarity, below the price cutoff, to be considered.

The regulations regarding water quality improvements were due to change during the lifecycle of the Tolt treatment facility. To account for this, the SPU requested designs for two possible options: one to cover the existing regulation (A) and one to cover the proposed improvement to that regulation (B). Both options were required to meet the minimum obligations for redundancy, environmental compatibility, efficiency, flexibility, longevity, and safety.

The RFP (see Exhibit 15-5 on the companion CD) required that buildings be designed for a fifty-year life and that the entire system meet the following flow requirements, options A&B:

- Minimum flow of 8mgd.

- Average flow of 120mgd (from May to October).

- Maximum flow, in the initial design, of 135mgd.

- Expansion capacity to 240mgd (plans for expansion must maintain the quality and standards set forth in the RFP and fit with the overall character of the site).

- Average flow rates must be maintainable with normal servicing underway.

- No facility shall include as processes: diatomaceous earth filtration, cartridge filtration or chlorine dioxide.

The RFP also required that no facility could go online until it passes an approved Acceptance Test where the facility will be operated at 135 mgd for 24 hours and between 8 and 120 mgd for 14 days. The 14-day test period shall include simulated events expected by the facility: power failure, regular maintenance, automatic shutdown and the full range of raw water conditions.

Exhibit 15-8   Request for Proposal (RFP)

The requirements to be met by the "B" option, involve a total 5-log removal of offending organisms for compliance, primarily Cryptosporidium. Because the requirements regarding Giardia and Cryptosporidium will eventually become law, the plan that met the "A" standard (which required a 4-log removal) included provisions for future adaptation to the "B" standard.

The proposing companies also needed to also meet the financial requirements of the SPU, including performance bonds and letters of credit from a sound financial backer of sufficient size to accommodate the project and see it to completion. All proposing teams received an honorarium of $100,000 for submitting a fully responsive proposal.

Evaluation of Proposals

Proposals were evaluated based on the proposing company's ability to meet the performance requirements of the RFP and the Service Agreement. Among the factors scored were:

1. Financial Criteria:
   - Cost Effectiveness.
   - Proposer's Financial Qualifications (including guarantor).

2. Team/Technical Criteria.
   - Project Implementability.
   - Technical Reliability.
   - Technical Viability.
   - Environmental Concern.
   - Proposer and Staff Past Performance.
   - WMBE (Women's and minority business usage requirement).

Evaluation was based on the following points system:
Financial Criteria................40 points
Team/Technical Criteria........60 points
           Total...............100 points

CDM Phillip's Winning Proposal

The winning proposal came from CDM Philip (see Exhibits 15-6 and 15-7). Like many of the other consortia formed to bid for this job, CDM Philip was a combination of companies: Philip Utilities Management Corporation, a subsidiary of Philip Environmental, Camp, Dresser and McKee Inc., and Dillingham Construction N.A. Inc. Philip Environmental Inc. is a publicly traded Canadian company that provides a wide range of environmental services, including management, to municipalities and industry. Camp, Dresser and McKee Inc. celebrated their 50th anniversary in 1997, and is a recognized leader in environmental engineering. These two companies joined forces with Dillingham Construction N.A. Inc., a nearly

120 year old construction firm that began by building Pearl Harbor in the late 1800's and laying the first of California's paved highways in 1920.

Exhibit 15-9   *CDM Philip Drawings*
Exhibit 15-10   *CDM Philip Spreadsheets*

Because the group's A proposal would require construction modifications in excess of 13 million dollars and a risk of service disruption, SPU decided to seriously consider Proposal B. The primary difference between the two proposals was the addition of ozone treatment in Proposal B. The ozone facilities and the ozone itself account for the majority of the construction and O&M price differences between the two proposals

The CDM Philip design differed significantly from the Benchmark design in three ways: the overall facility was more compact; gravity and hydraulics powered the primary means of solid removal; and the filter media were far less thick than the media proposed in the Benchmark.

In CDM Philip's design, the untreated water enters the system and first contacts ozone as it flows under and over barriers in a vertical "S" path. This vertical mixing maximizes contact with the ozone. Next the water passes a flash mixer, which is a jet of recirculated water that gives sufficient local turbulence to mix the chemicals added at that point. Just past the flash mixer is the flocculation tank. This is a series of vertical baffles that force the water to follow a horizontal "S" shaped path. This 180-degree turning of the water flow provides enough mixing to cause the smaller particles to collide and "floc" together, aided by the aluminum added at the flash mixing point. The compactness of the flocculation tanks and the turbulence created by the twisting water flow allow CDM Philip to avoid using mechanical mixing. Mechanical mixers add a significant level of complexity and expense, not only to plant construction but also to operations and maintenance.

Once past the flocculation tanks, the water flows through the filter beds. Here, CDM Philip specified 56 inches of carbon to filter the partially treated water. The Benchmark design included 80 inches of carbon and 10 inches of sand. After passing through the filters, the water moved to the massive Clearwells buried near the treatment plant. The water receives chlorine treatment while en route to the Clearwells.

As in the Benchmark, the Clearwell serves several purposes. First, the large volume of water, mixed by the incoming flow, helps to balance the quality of the final product. Because the water flowing from the Tolt River has seasonal variations in turbidity and chemical makeup, the Clearwell allows the different waters to mix before traveling out to the public. Second, the holding time allows the water to fully mix with the chlorine, providing adequate disinfection. When the water leaves the Clearwell, the final product is treated with pH balancing chemicals to help reduce corrosion in the pipes.

CDM Philip's two proposals, A & B, both showed more than 15% cost savings compared to the Benchmark. Proposal A cost just over $55 million to build and just under $1.9 million a year to operate. Proposal B cost just over $68 million to build and just under $2 million a year to operate. Proposal B, at an NPV just over $66 million for construction, undercuts the Benchmark by over 30% and the operations expenses undercut the Benchmark by over 20%.

Questions

1. What role did transparency play in the success of this procurement?

2. In July 2000, The American Bar Association adopted Revisions to the 1979 Model Procurement Code. The ABA 2000 Model Procurement Code authorizes simultaneous use of DBB, DB, DBO, and DBFO (see Exhibit 15-8 on companion CD). Does the new ABA Code provide a ready solution for Deborah as to scope of work, transparency, engineering safety confirmed, open to innovation, and sound financial plan?

Exhibit 15-11 Excerpts from the American Bar Association 2000 Model Procurement Code

3. SPU employed an independent design team to establish a clear baseline for potential competitors before the RFP was issued and to provide a clear picture of the evaluation process.   Evaluate this Benchmarking strategy.   What are its advantages and disadvantages to the city, proposers, and consultants? Include short-term, long-term and safety considerations.

4. The SPU's choice of discount rate and Construction and Operation Price Indices has a significant impact on the submitted cash flows in the proposals. Review the spreadsheets in Exhibit 15-2 to consider the following questions. Were their rates reasonable and realistic? What effect did these rates have on the proposals? How would a change in these rates affect the cash flows?

5. CDM Philip captured this project based partly on the strength of their knowledge concerning advanced filter technology. The filters proposed for use at Tolt are designed for a significantly higher flow rate than those specified in the Benchmark, and with a substantial savings in energy costs. What safeguards are in the RFP to protect against the possibility that the filters will require two, three even four times the cleaning and maintenance? How is the risk of failure of the technology handled?

References

Seattle Water Department (now Seattle Public Utilities, SPU). Tolt Treatment Facilities Request for Proposals Volumes I II & III: November 14, 1996.

CDM Philip Inc. Proposal – Seattle Water Department Tolt Treatment Facilities; Vols. I, II, & III: November 27, 1996.

S.P.U., Malcolm Pirnie, Inc., CDM Philip, and The University of Washington Institute for Public Policy and Management. The binder from: Water Utility Efficiency Workshop: Achieving Effectiveness Through Creative Change; The Westin Seattle Hotel – Seattle WA. November 20-22, 1997.

Notes

---

[36] <u>Background Issue Paper as Prepared for City Council</u>, November 1995.

About the Accompanying CD-ROM

Readers are encouraged to use the Excel® and Adobe® Acrobat® files enclosed on the CD-ROM to gain a better understanding of the cases presented in this book. Many of these files are required to be used to solve the questions presented at the end of each chapter. These files are included on the CD-ROM as an instructional aid.

The CD-ROM is distributed by Kluwer Academic Publishers with absolutely no support and no warranty from Kluwer Academic Publishers. Use or reproduction of the information on the CD-ROM for commercial gain is strictly prohibited. Kluwer Academic Publishers shall not be liable for damage in connection with, or arising out of, the furnishing, performance or use of the CD-ROM.

Excel® is a registered trademark of Microsoft Corporation. Adobe® and Acrobat® are registered trademarks of Adobe Systems Incorporated.